Gaining the edge in ruminant production

Gaining the edge in ruminant production

Nutritional strategies for optimal productivity and efficiency

edited by:

Sylvie Andrieu

Wageningen Academic
P u b l i s h e r s

ISBN: 978-90-8686-044-9

First published, 2007

Wageningen Academic Publishers
The Netherlands, 2007

Contents

Proteomics, nutrigenomics, metabolomics: do we have effective tools to optimise rumen function?

Karl Dawson
Director of Worldwide Research, Alltech, USA

1. Introduction

It has long been the dream of nutritionists to improve the efficiency and growth rate of livestock through careful manipulation of diets and management practices. This dream became a reality with the advent of the use of subtherapeutic antibiotics as growth promoters and growth permitants in livestock production systems. Even though these strategies are commonly used in modern livestock production systems, the underlying mechanisms that accounted for the beneficial effects of antimicrobial growth promoters have yet to be clearly defined. However, many new tools are now becoming available for defining the specific effects of growth promoting strategies at a very basic molecular level. These tools promise to better define nutritional and supplementation schemes and will have a tremendous impact on livestock production.

Over the last five decades, biologists have focused on developing an understanding of the basic genetic and molecular structures that describe life. This process started with early attempts to define the structure of the DNA molecule in the 1950's and ended in 2003 with the successful description of the complete nucleotide sequence that describes the human genome. This has been a very successful scientific era and has resulted in an understanding of both the basic physical and chemical nature of genes. The description of the basic nucleic acid sequences in the human genome has been a monumental accomplishment that has resulted in many new biochemical and computational technologies that have given modern biologists many exciting analytical tools. Using these tools, scientists have now described the complete sequence of information in the genetic codes of at least 100 different organisms. However, having completed these basic genomic projects, it is time to learn how to use this information. We are about to enter into a new era in biology. Efforts during this post-genomic era will focus on the very basic processes that regulate the genes and their expression, and will ultimately provide a quantitative description of all of the processes that sustain life.

2. Understanding antimicrobial growth promoters

Antimicrobial agents are often used in ruminant feeds to improve production efficiencies. In many instances, these supplements appear to alter ruminal fermentations by inhibiting specific groups of bacteria or specific metabolic processes. The most studied group of antimicrobial agents used in ruminants is the ionophores. These compounds are routinely used to specifically manipulate ruminal fermentations and to improve the performance of cattle by altering feed intake, increasing the efficiency of feed conversion, and increasing the rate of gain (Goodrich *et al.*, 1984). They are commonly used in feedlot diets, but may also be useful in improving the performance of dairy cattle. Numerous studies have evaluated the effects of the ionophores, monensin, and lasalocid, on microbial activities in the rumen (Table 1), and a great deal is known about their general mode of action.

Ionophore supplementation provides an excellent model of a successful method for improving the efficiency of animal production by manipulating the microbial activities in the rumen. It is generally believed that much of the growth-promoting activities of these compounds can be explained by their

Table 1. Some effects of ionophores on ruminal bacteria and fermentations parameters.

Item	References
Altered VFA patterns (enhanced propionate formation; decreased hydrogen, acetate, and lactate formation)	Richardson *et al.*, 1976; Chen and Wolin, 1979
Decreased methane production	Thornton *et al.*, 1976; Dinius *et al.*, 1976; Rumpler *et al.*, 1986
Decreased ruminal ammonia concentration	Dinius *et al.*, 1976
Decreased urease activities	Starnes *et al.*, 1984
Decreased proteolysis	Hanson and Klopfenstein, 1979
Decreased deamination	Van Nevel and Demeyer, 1977
Decreased rate of passage	Lemenager *et al.*, 1978
Decreased ruminal potassium and calcium concentrations	Starnes *et al.*, 1984
Inhibited growth of specific groups of bacteria	Chen and Wolin, 1979; Dennis *et al.*, 1981; Henderson *et al.*, 1981

ability to increase the energetic efficiency of the ruminal fermentation (Nagaraja *et al.*, 1997; Schelling, 1984). Ionophore-induced fermentation changes that result in increased formation of propionate and decreased acetate formation that can be stoichiometrically related to decreased methane formation. Since methane cannot be used as an energy source by the animal, such shifts in fermentation would be expected to provide more metabolisable energy to the animal for production. Other studies have suggested that ionophores can also exert a protein-sparing effect and can alter nitrogen metabolism in the rumen by inhibiting deaminase activity (Van Nevel and Demeyer, 1977) and decrease the activities of amino acid-degrading organisms (Chen and Russell, 1988; Russell *et al.*, 1991). Such activities can beneficially influence protein metabolism in ruminants. Other advantages associated with the use of some ionophores include decreased incidence of ruminal bloat (Bartley *et al.*, 1983) and acute bovine pulmonary edema and emphysema (Nocerini *et al.*, 1985).

Many nutritionists feel that our understanding of the effects of ionophores on ruminal fermentations is complete and easily defined. However, most microbiologists feel there is a great deal to learn about the overall effects of these materials on ruminal microbial populations and individual groups of ruminal microorganisms. It is generally believed that the action of ionophores in the rumen can be explained by their selective antimicrobial activities in the rumen (Chen and Wolin, 1979). These activities inhibit the growth of bacteria that contribute to the processes leading to methane formation and protein degradation while allowing for the growth of the organisms to produce end products that are more easily used by the animal. Recent studies have suggested that these selective antimicrobial activities may not completely explain the effects of ionophores on ruminal metabolism, because it has not always been possible to associate increased prevalence of ionophore-resistant bacteria with beneficial changes in ruminal fermentations (Dawson and Boling, 1983). Other studies with strains of bacteria that have been selected for resistance to the ionophore, monensin, suggest that these antimicrobials are capable of inducing specific metabolic changes within certain types of bacteria (Morehead and Dawson, 1992). Such changes result in alterations in the metabolic pathways used for energy production within the bacterial cells and suggest that ionophores can regulate microbial activities at a basic metabolic level, rather than by simply inhibiting specific groups of bacteria. In the future, a complete understanding of these basic metabolic shifts may suggest alternate ways of manipulating energy metabolism in the rumen that are not dependent on the complex action of ionophores. However, advances in this area await new techniques for evaluating the physiological effects of ionophores.

Unfortunately, a clear understanding of the mechanisms that explain the effects of many of the other antimicrobial growth promoters used in livestock production is still lacking. While it is clear that the common antimicrobial activities are common characteristics of many growth promoters, there are still a large number of proposed but unproven mechanisms that are used to explain their ultimate effects on animal performance. These include control of pathogens that cause subclinical infections, suppression of toxin formation, nutrient-sparing effects, the production of microbial-induced growth factors, and alterations in gut development patterns. In addition, it is likely that these supplements influence the development of the immune system and may significantly alter the amounts of energy needed to maintain a healthy gastrointestinal tract. Each of these hypothetical mechanisms has merit, but, to date, no single mechanism can be used to explain all of the effects of antimicrobial compounds used as growth promoters. With the advent of new governmental regulatory restrictions on the use of antimicrobial growth promoters, there is an increased need for a new approach to evaluate their mode of action, so that alternative growth promotion strategies can be defined.

3. Using genomic information for understanding growth promotion

Genomics is the study of genetic structures at the very basic molecular level and their relation to basic biological processes. In recent years, genomics has fostered the development of a multitude of new disciplines based on some basic molecular tools that have developed as we improve our knowledge of the basic molecular structure of life. Sciences like transcriptomics, proteomics, and metabolomics are being developed to study the quantitative relationships between the gene expression, protein production, and metabolic processes, respectively (Figure 1). These sciences are being combined to better describe the basic flow and regulation of all metabolic functions and can be used to explain the effects of nutrition and management practices on livestock growth, performance, and health.

At its very basic level, biology can be defined by a central dogma that describes the flow of information from DNA to RNA to protein (Figure 1). It is the regulation or control of the information flow in this pathway that delineates all biological processes from the basic development of cell structure to the most complex expressions of immune function. Many factors in the environment facilitate this regulation. These include disease challenge, exposure to environmental toxins, and nutrient supplies. It has long been the goal of both basic scientists and production agriculturalists to develop schemes

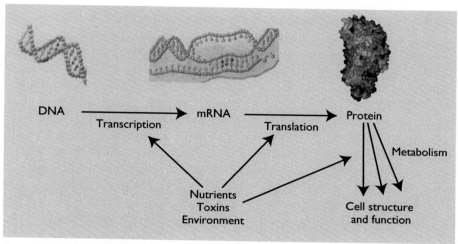

Figure 1. Regulation of information flow in biological systems.

for understanding and controlling the flow of this information. Our ability to understand these regulatory processes has changed considerably with the delineation of the various animal, plant, and microbial genomes. One step in this process, transcription, can now be described in detail by looking at the expression of various genes. This particular step is the first point in the regulatory processes, which controls the flow of information from the base genes to the development of the proteins, and, in turn, defines all metabolic processes and cellular structure. An understanding of these gene expression patterns in the presence of growth promoting antibiotics will help us better define the molecular basis for improved animal performance. While we have had techniques to study the expression of individual genes for many years, we now have tools that will allow us to do this on an unprecedented scale. The information obtained from genetic sequencing studies can be used to specifically describe the regulation of thousands of genes at the same time.

While it is clear that the growth promoters used in animal agriculture can influence the basic molecular mechanisms in the cells, we currently have very little information about the specific processes that are influenced at the transcriptional level. New molecular tools will be useful in helping us develop this understanding.

4. Description of microarrays as a tool for studying ruminants

In the last five years, our knowledge of nucleic acid sequences have provided tools that can be used to get a clearer understanding of overall gene expression at the transcriptional level. While these studies of gene transcription have many drawbacks, our ability to globally evaluate gene expression provides us with many tools for elucidating the key processes in metabolic regulation. We now have powerful screening methods to identify the key gene expression patterns that are influenced by environment and nutrition, and these methods will be critical in developing an understanding of growth promotion strategies. The methods are tools that have never been available before the completion of the various genome projects and promise to provide us with the information we need to develop alternatives to the commonly used antimicrobial growth promoters.

Oligonucleotide-based or cDNA microarray technologies allow for large-scale screening of gene expression at the level of transcription. These techniques are based on a quantitative assay of messenger RNA (mRNA) produced during the transcription of specific genes (Figure 2). The amount of the RNA transcript present in tissues reflects gene expression and can be measured indirectly after creating a complimentary labelled strand of DNA. This labelled material can be reacted or hybridised with an array of a known set of gene sequences that are attached to a solid glass slide or nylon substrate. The sequences are often organised as an array of small spots on the solid matrix and are generally referred to as probes. The intensity of the colour that results during the hybridisation process is directly related to the amount of target mRNA produced and reflects the level of gene expression. In this way, it is possible to determine which genes are upregulated (turned on) or downregulated (turned off) as a result of specific biological manipulations. Comparison studies of gene regulation can be carried out using contrasting colour labels on complimentary DNA from two sets of messengers. As a result, it is possible to quantitatively compare gene expression in two contrasting groups of tissue or animals. By using robotic techniques for producing arrays on a minute scale and laser techniques to discern the colour of specific spots, it is possible to examine the expression of thousands of genes at one time. This is an extremely powerful tool that can be used to study metabolic processes at a very basic level and lends itself well to understanding interactions that regulate gene function.

In the past, microarray studies have depended on specific arrays with relatively few nucleotides and limited amounts of information. These arrays are generated

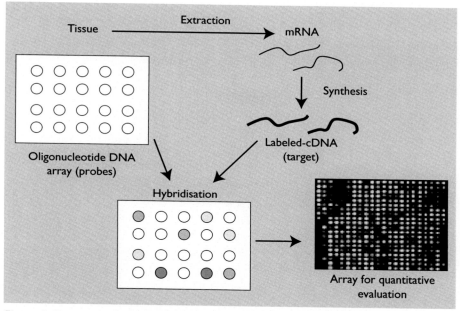

Figure 2. Preparation of an oligonucleotide microarray for quantitative evaluation of gene expression.

to examine specific metabolic functions or immune responses. This is now changing. Recent work has reported the development of array, which can be used to examine gene expression in bovine tissues. While the use of smaller, more defined arrays to examine regulation of specific tissue response have been useful, the development of standardised systems for examining the expression of large numbers of genes will greatly enhance our ability to understand basic metabolic and physiological functions in cattle.

5. Use of microarray technologies to understand gene regulation and growth promotion

Microarray studies for studying gene expression in cattle are still relatively rare. However, a variety of microarray-based studies have been completed or are underway in cattle (Table 2). Most of these studies focus on changes in gene expression that are associated with specific immunological responses, but others have focused on tissue development during growth. The ability of these techniques to simultaneously and efficiently examine the expression of thousands of genes at one time makes these techniques extremely powerful

Table 2. Some uses of oligo-based microarrays to study gene expression in cattle.

Topic	Reference
Identification of genes associated with immune suppression and mastitis susceptibility	Burton *et al.*, 2001
Defining gene expression patterns in bovine neutrophils during parturition	Madsen *et al.*, 2004
Examining effects of pregnancy on the expression of genes in the bovine placenta and uterus	Ishiwata *et al.*, 2003
Define gene expression changes associated with consumption of endophyte-infected tall fescue diets	Jones *et al.*, 2004
Evaluation of gene expression in cattle fed high and low quality rations	Reverter *et al.*, 2003
Evaluation of gene expression in cattle during nutritional restriction	Byrne *et al.*, 2005

and allows us to examine complex interactions, such as those associated with growth promotion.

Using these techniques, the relationships between numerous genes can be evaluated in a relatively short time, can be used to define multiple gene interactions, and will soon give us the screening tools we need to clearly define the growth promotion process at its most basic level. These techniques go far beyond traditional approaches that simply looked for phenotypic changes, changes in animal health, or changes in growth and efficiency. Studies using such techniques can be used to quantitatively look at the basic regulation of metabolic processes without having to carry our long-term growth studies and give many clues about expected physiological responses that, until now, were unknown. However, from a production point of view, the greatest value of techniques for measuring gene expression may be in their ability to elucidate the effects of critical management practices on metabolism and physiological responses. In this respect, they provide tools for nutritionists and production scientists that will allow for a cleaner control of the complex genetic interactions that drive all aspects of growth promotion. In mammalian systems, microarray technologies are being used to couple other molecular techniques to help define how nutrition and management influence the animal's physiology, metabolism, and the production responses.

6. Nutrigenomics

While there are, and will continue to be, many consequences of modern genomic research, the potential for understanding the interactions between genes and nutrition will undoubtedly provide some of the most powerful tools for modern dairy and beef production systems. While still in its infancy, the science of nutrigenomics strives to provide a molecular understanding of how common dietary components and nutrition affect animal growth and health by altering the gene expression or the phenotypic manifestation of an individual's genetic makeup. The goal is to provide nutritional or management strategies for control of processes associated with gene expression. In human medicine, nutrigenomics is being used to help develop designer or prescription diets that address individual genetic deficiencies, to address specific genetic diseases, and to overcome metabolic changes that are associated with aging or cancer. There are a number of basic tenets that provide the basis for this new branch of biology:

1. Common dietary chemicals act on the genome to alter gene expression or structure.
2. Under many circumstances, diet/gene interactions critically influence tissue structure and function.
3. An individual's or population's genetic make-up will influence responses to dietary manipulation.

As this science develops, it will provide a clear definition of the important genes involved with all aspects of development, growth, and health. More importantly, this science promises to define new nutritional strategies and management tools for altering gene functions that lead to improved animal health and productivity. When coupled with new genetic selection strategies, we can expect a significant impact on the way we view livestock production and management.

Nutrigenomic studies in beef and dairy cattle are still rare, but they are becoming important as we develop an understanding of the relationship between nutrition, genetics, and tissue growth. In studies of steers under nutritional restriction due to intake of poor quality feeds, expression of genes associated with protein turnover, cytoskeletal remodelling, and metabolic homeostasis was clearly influenced by diet (Byrne, *et al.*, 2005). While there is still much to be learned about the effects of overall gene expression on animal performance and the production of high quality food products, we now have specific tools to examine this process at the molecular level and relate important performance parameters to basic metabolic processes.

7. Effects of selenium on gene expression

Currently, only a few studies have examined the effects of nutrition on gene expression in cattle (Table 3). It is often taken for granted that these interactions exist, but the complexity of the interactions has made them difficult to elucidate. This is a frontier that will be addressed as we take advantage of the newly available information on the genetic sequences obtained from bovine genome studies.

A number of recent studies in mammalian systems have demonstrated the power of such gene expression studies in mammalian species (Swanson *et al.*, 2003). A good example is the recent work that has examined the effects of dietary selenium on gene expression in mice (Rao *et al.*, 2001). In studies using a 12,000-gene mouse microarray, it has clearly been established that selenium deficiencies can influence the patterns of protein synthesis in mice, and that these changes are partially regulated at the level of gene transcription (Table 3). It is not surprising that these changes in gene expression can be used to account for many of the outward characteristics of selenium deficiencies.

It is clear in these studies with mice that low selenium diets induced stress responses at the transcriptional level, and that low levels of dietary selenium can play a key regulatory role in many tissues. Such responses result in the increased oxidative stress and DNA damage associated with reduced production in selenoproteins and detoxification enzymes. The induction of genes involved in the regulation of cell cycling and oncognesis also explains the increased potential for uncontrolled cellular proliferation and increased potential for tumor development. However, the more important observation from these studies is that the intake of a single dietary micronutrient can predictably influence gene expression at a very basic level.

From a practical point of view, these molecular observations can explain how a single nutrient, selenium, can reduce cancer risks in humans by regulating basic gene expression. Again, all of these observations are the result of an enhanced understanding of the basic structural characteristics of the DNA that have been developed during these genome projects.

Table 3. Examples of major genes expression influenced by selenium deficiency in mice (Adapted from Rao et al.*, 2001).*

Gene group	Genes which were most significantly influenced	Specific function influenced
Upregulated genes		
Stress response genes	W41070, AA008244, X54149, X51829, AA061016, W30116	Genes associated with DNA repair and energy homeostasis
Cell growth and cycle control genes	AA020101, X66032, U06924, U00454, U77630, X067620	Genes associated with cell cycle, development and tumor development
Cell adhesion and angiogenesis genes	AA03323, U43298, M28730, U43836, M90365, W34697	Genes associated with membrane attachment, vascular development and specific cytoskeletal proteins
Downregulated genes		
Detoxification genes	M60273	Cytochromes responsible for xenobiotic metabolism
Selenoproteins	AA038494	Glutathione peroxidase enzyme for peroxide degradation
Lipid transport genes	AA087320, M64250, W17412, W62976	Genes for lipid transport intracellularly and in plasma
Cell adhesion and angiogenesis genes	M62860, W16201	Proteins involved in membrane adhesion and morphological signalling
Growth control	U30482, D12885	Hormonal regulation and hormone receptor proteins

Other nutritional studies have examined the effects of caloric intake on gene expression during the aging process (Lee *et al.*, 2002). In these studies, it was observed that the expression of more than 21% of the heart tissue transcripts in one 9,977-gene mouse array were altered by dietary caloric restriction while the aging process itself altered the expression of only 10% of the genes in the array. Such studies clearly demonstrate that in some tissue, the nutritional environment can have a greater impact on gene expression than the normal regulatory processes associated with aging.

The results from these mouse studies clearly demonstrate how an individual nutrient, or class of nutrients, can have broad effects on gene expression. This type of data provides critical information and new techniques for evaluating nutrient effects. In addition, these initial studies provide key information about specific marker or biochemical pathways that can be used to monitor nutritional effects on individual dietary components and relate to the role of nutrients in regulating animal health and development. It is very likely that similar changes in gene expression will be observed in cattle and can be used to define induced growth promotion patterns.

In the long term, it is likely that transcriptional profiles can provide a number of very basic tools for evaluating the nutritional and physiological status of an animal or group of animals. This could also be used to provide a detailed analysis of growth promotion strategies and to define alternative methods for enhancing animal performance without using antimicrobial growth promoters. The days of simply looking for deficiency symptoms or waiting for growth responses to understand basic nutrition and management strategies will pass into history. However, it can easily be seen from these types of gene expression studies that nutritional/gene interactions are extremely complex. When this is coupled with the differences in gene expression associated with the various critical tissues of the body and the potential for changes associated with age and developmental patterns, it is clear that the true challenges in the future will be in modelling and interpreting the massive amounts of information obtained from simple nutritional studies. This will challenge our computational abilities and will lead to a further merger of computer sciences and molecular biology. While these challenges seem daunting, they are no more formidable than those faced by the pioneer researchers who set out to describe the complexity of the functional structure of DNA that led to the full description of the genome 50 years ago.

8. Concluding remarks

The use of oligonucleotide-based arrays and similar molecular techniques to define transcriptional responses to nutritional management manipulations has tremendous potential. The resulting gene expression profiles can be useful in understanding the basic regulation of metabolism as it is controlled at the molecular level, and will be extremely useful for the development of novel nutritional and management strategies. It will very likely provide the tools necessary to understand growth promotion strategies. As compared to currently used production responses, this work can identify the key metabolic parameters that are readily influenced by current growth promoting strategies and to define new strategies. Such technology can be used to take the guesswork out of nutritional and management studies that rely on rather subtle changes in often difficult-to-measure response criteria.

References

Bartley, E.E., T.G. Nagaraja, E.S. Pressman, A.D. Dayton, M.P. Katz and L.R. Fina, 1983. Effects of lasalocid or monensin on legume or grain (feedlot) bloat. *J. Anim. Sci.* **56**:1400-1406.

Burton J.L, S.A. Madsen, J. Yao, S.S. Sipkovsky and P.M. Coussens, 2001. An immunogenomics approach to understanding periparturient immunosuppression and mastitis susceptibility in dairy cattle. *Acta. Vet. Scand.* **42**:407-427.

Byrne, K.A., Y.H. Wang, S.A. Lehnert, G.S. Harper, S.M. McWilliam, H.L. Bruce and A. Reverter, 2005. Gene expression profiling of muscle tissue in Brahman steers during nutritional restriction. *J. Anim. Sci.* **83**:1-12.

Chen, G. and J.B. Russell, 1988. Fermentation of peptides and amino acids by a monensin-sensitive ruminal peptostreptococcus. *Appl. Environ. Microbiol.* **54**:2742-2749.

Chen, M., and M.J. Wolin, 1979. Effect of monensin and lasalocid-sodium on the growth of methanogenic and rumen saccharolytic bacteria. *Appl. Environ. Microbiol.* **32**:703-710.

Dawson, K.A. and J.A. Boling, 1983. Monensin-resistant bacteria in the rumens of calves on monensin-containing and unmedicated diets. *Appl. Environ. Microbiol.* **46**:160-164.

Dennis, S.T., T.G. Nagaraja, and E.E. Bartley, 1981. Effects of lasalocid or monensin on lactate-producing or -using rumen bacteria. *J. Anim. Sci.* **25**:418-426.

Dinius, D.A., M.E. Simpson, and P.B. Marsh, 1976. Effects of monensin fed with forage on digestion and the ruminal ecosystem of steers. *J. Anim. Sci.* **42**:229-234.

Goodrich, R.D., J.E. Garrett, D.R. Gast, M.A. Kirck, D.A. Larson and J.C. Meiske, 1984. Influence of monensin on the performance of cattle. *J. Anim. Sci.* **58**:1484.

Hanson, T.L. and T. Klopfenstein, 1979. Monensin, protein source and protein levels for growing steers. *J. Anim. Sci.* **48**:474-479.

Henderson, C., C.S. Stewart and F.V. Nekrep, 1981. The effect of monensin on pure and mixed cultures of rumen bacteria. *J. Appl. Bacteriol.* **51**:159-169.

Ishiwata, H., S. Katsuma, K. Kizaki, O.V. Patel, H. Nakano, T. Takahashi, K. Imai, A. Hirasawa, S. Shiojima, H. Ikawa, Y. Suzuki, G. Tsujimoto, Y. Izaike, J. Todoroki and K. Hashizume, 2003. Characterization of gene expression profiles in early bovine pregnancy using a custom cDNA microarray. *Mol. Reprod. Dev.* **65**:8-18.

Jones, K.L., S.S. King and M.J. Iqbal, 2004. Endophyte-infected tall fescue diet alters gene expression in heifer luteal tissue as revealed by interspecies microarray analysis. *Mol. Repro. Dev.* **67**:154-161.

Lemenager, R.P., F.N. Owens, B.J. Shockey, K.S. Lusby and R. Totusek, 1978. Monensin effects on rumen turnover rate, twenty-four hour VFA pattern, nitrogen components and cellulose disappearance. *J. Anim. Sci.* **47**:255-261.

Lee, C.K., D.B. Allison, J. Brand, R. Weindruch and T.A. Prolla, 2002. Transcriptional profiles associated with aging and middle age-onset caloric restriction in the mouse heart. *PNAS.* **99**:14988-14993 (www.pnas.org/cgi/doi/10.1073/pnas.232308999)

Madsen, S.A., L.C. Chang, M.C. Hickey, G.J.M. Rosa, P.M. Coussens and J.L. Burton, 2004. Microarray analysis of gene expression in blood neutrophils of parturient cows. *Phyisol. Genomics* **16**:212-221.

Morehead, M.C. and K.A. Dawson, 1992. Some growth and metabolic characteristics of monensin-sensitive and monensin-resistant strains of *Prevotella (Bacteroides) ruminicola. Appl. Environ. Microbiol.* **58**:1617-1623.

Nagaraja, T.G., C.J. Newbold, C.J. VanNevel and D.I. Demeyer, 1997. Manipulation of ruminal fermentation. In: *The rumen microbial ecosystem* (2nd Ed.). Chapman and Hall. pp. 523-632.

Nocerini, M.R., D.C. Honeyfield, D.C. Carlson and R.G. Breeze, 1985. Reduction of 3-methylindole production and prevention of acute bovine pulmonary edema and emphysema with lasalocid. *J. Anim. Sci.* **60**:232-238.

Rao, L., B. Puschner and T.A. Prolla, 2001. Gene expression profiling of low selenium status in the mouse intestine: Transcriptional activation of genes linked to DNA damage, cell cycle control and oxidative stress. *J. Nutr.* **131**:3175-3181.

Reverter, A., K.A. Byrne, H.L. Bruce, Y.H. Wang, B.P. Dalrymple and S.A. Lehnert, 2003. A mixture model-based cluster analysis of DNA microarray gene expression data on Brahman and Brahman composite steers fed high-, medium- and low-quality diets. *J. Anim. Sci.* **81**:1900-1910.

Richardson, L.F., A.P. Raun, E.L. Potter, C.O. Cooley and R.P. Rathmacher. 1976. Effect of monensin on rumen fermentation in vitro and in vivo. *J. Anim. Sci.* **43**:657-664.

Rumpler, W.V., D.E. Johnson and D.G. Bates, 1986. The effect of high cation concentration on methanogenesis by steers fed diets with and without ionophores. *J. Anim. Sci.* **62**:1737-1741.

Russell, J.B., R. Onodera and T. Hino, 1991. Ruminal protein fermentation: New perspectives on previous contradictions. In T. Tsuda, Y. Sasaki and R. Kawashima (Ed.). *Physiological Aspects of Digestion and Metabolism in Ruminants.* p. 681. London Academic Press.

Schelling, G.T., 1984. Monensin mode of action in the rumen. *J. Anim. Sci.* **58**:1518.

Starnes, S.R., J.W. Spears, M.A. Froetschel and W.J. Croom, 1984. Influence of monensin and lasalocid on mineral metabolism and ruminal urease activity in steers. *J. Nutr.* **114**:518-525.

Swanson, K.S., L.B. Schook and G.C. Fahey, 2003. Nutritional genomics: Implications for companion animals. *J. Nutr.* **133**:3033-3040.

Thornton, J.H., F.N. Owens, R.P Lemenger and P. Totusek, 1976. Monensin and ruminal methane production. *J. Anim. Sci.* **43**:336.

Van Nevel, C.J. and D.I. Demeyer, 1977. Effect of monensin on rumen metabolism in vitro. *Appl. Environ Microbiol.* **34**:251-257.

Bio-based economy and new challenges to dairy nutrition

Ad van Vuuren
Animal Sciences Group of Wageningen UR, Lelystad, The Netherlands

1. The bio-based economy

Foods produced from dairy cow's milk are a major source of nutrition for many consumers, whether consumed as daily staples or luxury goods. Changes in consumer awareness have lead to increased demand for a plentiful supply and choice of dairy foods, which must be cheap and readily available. Purchasers expect their food to be safe, i.e. free of any disease-causing entities, and also be healthy in terms of the type of nutrients delivered to the consumer. To an increasing extent farmers are expected to produce milk in an energy efficiently manner as well as from sustainable resources, in this case feedstuffs. As a result, there has been a major shift from 'factory' farming to higher accountability in agriculture, with consumers expecting milk producers to embrace modern ideas regarding efficiency and sustainability of dairy production, especially with respect to the environment and, as the media term it, a low 'carbon footprint'.

A bio-based economy can be defined as a complex of economic activities using biomass, which, in turn, can be defined as organic, rather than fossil or mineral, material. Life on Earth is derived primarily via photosynthesis in plants, which convert sunlight into biomass. This biomass may be used as a food source for humans (in the case of crops) and for herbivorous animals producing meat or milk, or as a source of fuel (traditionally wood-derived). Global demands for energy are expected to increase to a total of 400×10^{18} Joules by the year 2050. Thirty percent of total demand will be for bio-diesel and bio-ethanol. Land areas, world-wide, devoted to the production of crops for bio-fuels are forecast to reach a total of 10% of all available land at this time.

Let's take a look at bio-fuels specifically. Bio-diesel is produced from oil-rich plant material, including soybeans, oil palms and sunflower seeds. There may also be opportunities in future for using animal fat as a source material. Co-products that result from manufacturing bio-diesel include grain or legume meals, expeller waste, kernels and glycerol.

Bio-ethanol is derived from materials rich in carbohydrates (starch and sugar). Sources used include maize, wheat, barley and rye, as well as sugar cane, beet and molasses. Co-products left over after processing of these sources include distiller's grains, sugar beet pulp and vinasses. Other 'bio-based' products include enzymes and probiotics from fermentation with either bacteria or fungi. By-products from such processing include the digested (whole or part) substrate and the microbial biomass.

The result of increased production of bio-fuel is likely to result in higher prices for starch-rich plant commodities, in conjunction with an increased variety and abundance of new feed materials as by-product from this fuel manufacturing. These materials are typically characterised by their low dry matter concentrations, high crude protein and mineral content and high variability in quality and consistency of nutritional profile.

2. Consequences for dairy cows diets: dietary starch supply

What would be the consequences of feeding such materials to dairy cows? An increased use of starch-for bio-ethanol production, reduces the availability of dietary starch, which forms the main energy source in a balanced ration for early-lactating cows. If starch levels are decreased, reduced levels of propionic acid from rumen fermentation (Table 1) and glucose absorption can be expected.

Glucose and propionic acid are glycogenic precursors required for the production of glucose and lactose. Pregnant cows require a good supply of amino acids and glucose for their developing foetus, and need increased storage of fatty acids in the adipose tissue as an energy reserve for the onset of lactation. Once lactation starts, protein, glucose and fatty acids are routed, to a large extent, to the udder in order to ensure adequate milk production. Following birth, fat is mobilised

Table 1. Percentage of each substrate fermented into the specific VFAs (Bannink et al., 2006).

Substrate	% of substrate fermented to specific VFA			
	Acetic	Propionic	Butyric	Others
Soluble sugars	53	16	26	6
Hemi-cellulose	51	12	32	5
Cellulose	68	12	20	1
Protein	44	18	17	21
Starch	49	31	15	5

and liberated via the liver to address the negative energy balance created in the dairy cow. As storage of glucogenic precursors is minimal in dairy cows, the hexose requirement for milk (lactose) production should be met mainly by the absorption of glucogenic precursors from the gastro-intestinal tract. It can be calculated that the production of 40 kg of milk (1.8 kg of lactose) requires 11 moles of hexoses. Assuming 0.4 kg of organic matter to be fermented into VFA per kg of dry matter, 19 Mol% of VFA as propionic acid, and 2 moles of propionic acid yielding 1 mol of hexose, it can be calculated that more than a dry matter intake of more than 30 kg is required to supply 11 moles of hexoses (Figure 1).

Extra starch in rations for early-lactating dairy cows will increase the proportion of substrate that will be fermented into propionic acid and in the case of slowly fermentable maize starch also increases the supply of by-pass starch that will be absorbed as glucose in the small intestine.

It is obvious that changes in starch availability for dairy cow rations, due to the introduction of bio-ethanol or bio-diesel by-products, will have to be compensated to maintain hexose supply and thus milk (lactose) production. Feed additives that have the potential to alter VFA proportion may have an

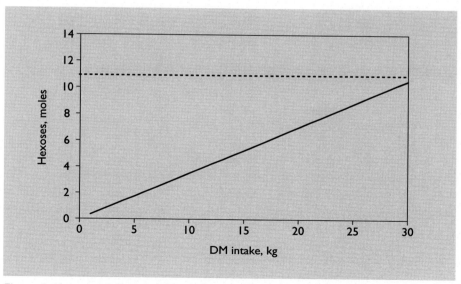

Figure 1. Hexose requirement for 40 kg of milk (doted line) and relationship between hexose supply and DM intake (solid line).

application, as well as glycerol, which the emerging bio-based economy will produce in large quantities. Changes in feed ingredients, such as the inclusion of glycerol, can alter the proportion of individual VFA produced, especially that of propionic acid as shown by DeFrain *et al.* (2004; Table 2). Although ruminal propionic acid was increased by adding glycerol in the diet of transition cows, postpartum concentrations of glucose in plasma were highest for cows fed the control diet relative to glycerol supplemented diets. Concentrations of beta-hydroxybutyrate and non–esterified fatty acids were similar among treatments. In this study, energy-corrected milk production tended to be lower in the glycerol supplemented diets relative to control diet. The effects on milk production still need to be explored, however, as research results have been inconsistent until now.

Table 2. Effect of glycerol on propionic acid and milk yield (DeFrain et al., 2004).

Parameter	Treatment		
	Control	Low glycerol	High glycerol
Dry matter intake, kg/d	17.9	17.5	15.8
Starch intake, kg/d	6.2	5.7	4.8
Glycerol intake, kg/d	0.0	0.4	0.9
Propionic acid, mol%	21.7***	27.1	24.7
Energy-corrected milk, kg/d	38.7*	35.2	35.0

*p<0.10; ***p<0.01.

3. Dietary protein supply

Bio-based manufacturing will also produce by-products rich in protein. Therefore it is likely that protein will no longer be the limiting factor of dairy cow diets, with farmers seeing an increase in dietary protein supply. This change will also affect other important considerations, including the excretion of nitrogen and phosphorous and the effects on efficiency of utilisation of nitrogen and phosphorous on dairy farms. Increasing nitrogen intake in dairy cows mainly results in an increase in urea nitrogen excretion by urine (Figure 2). The increased excretion of nitrogen will have a negative impact on nitrogen emissions from dairy farms. The increased availability of high-protein diets therefore pushes farmers to high-quality mineral management.

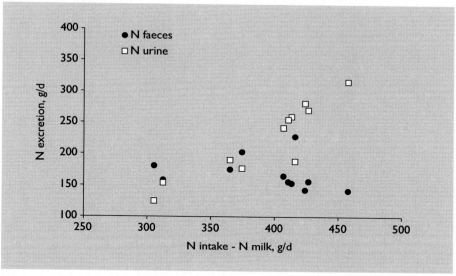

Figure 2. Effect of excessive nitrogen intake on nitrogen excretion rates in urine and faeces of dairy cows.

4. Feed dry matter

Transportation costs determine if bio-fuel by-products will be delivered in dry or wet form. Feeding these by-products with lower dry matter content than conventional dietary ingredients, may affect dry matter intake. Dry matter intake is a result of the intake capacity of the animal and the satiety value of a particular feed. For grass silage, Zom *et al.* (2002) observed an increase in the satiety value as dry matter concentration decreased (Figure 3). For maize silage, the satiety factor increased as dry matter concentration was lower or higher than 350 g/kg (Figure 4). Consequently, intake of grass and maize silage will be reduced at lower dry matter concentrations.

It is, however, uncertain if moisture content *per se* affects dry matter intake. In a study of De Visser and Hindle (1992), adding water to wilted silage had no effect on dry matter intake, whereas ensiling at a lower dry matter concentration with either molasses or formic acid, reduced dry matter intake (Table 3). These authors concluded that the increased concentration of fermentation products in the wet ensiled silages affected dry matter intake.

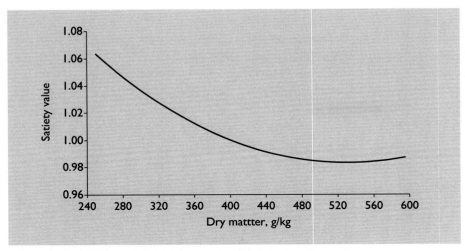

Figure 3. Effect on satiety factor for grass silage (Zom et al., 2002).

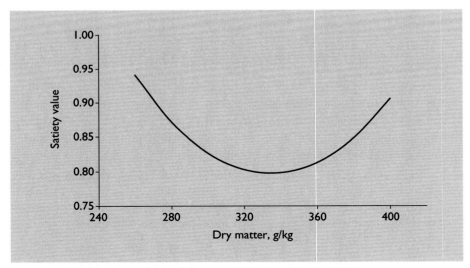

Figure 4. Effect on satiety factor for maize silage (Zom et al., 2002).

So, if dairy feeds are formulated with by-products with low dry matter concentration (and, therefore, higher moisture), there is a potential impact on total energy intake, depending on the ration dry matter concentration. It may also affect storage quality, with increased risk of effluent from the clamp and mould growth, the latter increasing mycotoxin levels and reducing

Table 3. Effect of TMR dry matter on milk production (De Visser and Hindle, 1992).

Parameter	Treatment (30% grass silage dry matter)			
	Wilted	Wet + molasses	Wet + formic acid	Wilted + water
TMR DM, %	39.9	33.6	33.1	33.6
DMI, kg	21.9[a]	19.6[b]	20.0[b]	21.0[a]
NEL, MJ	143[a]	130[a]	129[a]	137[a]
Milk yield, kg FPCM	38.0[a]	36.2[b]	36.2[b]	38.1[a]

[a,b]Means not sharing a letter differ significantly.

appetite. In this respect, an increased dependence on silage inoculants and storage protection ingredients will result, if milk production levels are to be maintained.

5. Improving feed quality

A heavier reliance on feed materials with poorer nutritional values can result in lower digestibility and energy supply to the animal. Pre-treatment with enzymes or even using pre-fermentation ingredients may become more important to maintain cow performance as these new feed materials have a relatively high lignin content.

The higher variability in nutritional content and quality of by-products used in cow rations will require new and more rapid analytical assessment of feeds, to ensure appropriate levels of nutrients are being delivered.

6. Conclusions

Consumers increasingly expect readily available sources of food and fuel. In the last couple of years, this has resulted in a strong demand for bio-fuels, which will inevitably change the availability of raw materials for use in animal feeds. There will be increased competition in world markets for plant-based materials to be used as food or fuel. This will mean that, for the ruminant feed formulation, less consistently high quality materials will be available for balancing rations. Less starch and more protein are likely to figure in the nutritional profile, which will alter the delivery of glucose and other energy sources as well as nitrogen excretion. Feedstuffs with a lower and more variable quality will require enzyme or other pro-nutrient treatments to minimise potential negative impacts on milk production. Also, low dry matter diets require more vigilance during storage, and should be treated to reduce the risk of mycotoxin contamination from mould growth.

References

Bannink, A., J. Kogut, J. Dijkstra, E. Kebreab, J. France, S. Tamminga and A.M. Van Vuuren. 2006. Estimation of the stoichiometry of volatile fatty acid production. *Journal of Theoretical Biology* **238**: 36-51.

DeFrain, J.M., A.R. Hippen, K.F. Kalscheur and P.W. Jardon. 2004. Feeding glycerol to transition dairy cows: effects on blood metabolites and lactation performance. *Journal of Dairy Science* **87**: 4195-4206.

De Visser, H. and V.A. Hindle. 1992. Autumn-cut Grass Silage as Roughage Component in Dairy Cow rations. I. Feed Intake, Digestibility and Milk Performance. *Journal of Agricultural Science* **40**: 147-158.

Zom, R.L.G., J.W. Van Riel, G. André and G. Van Duinkerken. 2002. *Voorspelling voeropname met Koemodel 2002* [Predicting feed intake with Cow Model 2002]. Praktijk Rapport Rundvee nr. 11. Praktijkonderzoek Veehouderij, Lelystad. 50 p.

Dairy nutrition models: their forms and applications

William Chalupa, Ray Boston and Robert Munson
School of Veterinary Medicine, University of Pennsylvania, Kennett Square, PA 19348, USA

1. Introduction

For some years it has been evident that dairy cow nutrition models are vital to the continued success of the dairy industry. This is especially true as we recognise the importance, for example, of ruminal microbes and metabolism in body tissues to nutrient requirements. In addition, our production emphasis has shifted from only milk volume and fat to include milk protein percentage and yield. Mathematical models of nutrition have been in use for over three decades and have stimulated improvement in feeding cattle. However, more complete data sets available in recent years combined with more precise mathematical approaches have now allowed us to improve models of nutrient use tremendously. Such models will be used more frequently in the future for support of decisions not only on the nutrition of cattle, but for other aspects including farm economics and environmental impact.

2. Dairy nutrition models: their form and role

Nutritional models vary in complexity according to objectives. A typical scheme of model levels needed to represent a system is found in Table 1. Information about a system must be at least one level below the system explored with the model. Thus, models describing herds operate at the animal level or below, those describing animals require details at the organ level and lower and so on.

Table 1. Model levels (Adapted From France and Thornley, 1984).

Level	Description of level
i + 1	Collection of organisms (herd, flock, crop)
i	Organism (animal, plant)
i-1	Organs
i-2	Tissues
	Cells
	Organelles

In practice, models only need details that have significant bearing on consequences of changes arising from inputs to the system (Production Model) or as much detail as is necessary to explore the system in new and different ways (Scientific Model). Salient properties of production and scientific models are presented in Table 2.

2.1. Scientific models

Scientific models are usually developed upward from basic experimental data pertaining to metabolic processes. Scientific models assume that a living system can be described in terms of a set of 'critical' metabolic transactions encapsulated in organs. The goal is to translate *in vitro* experimental data into chemical reactions representing the essential metabolic processes. Differential equations of the mass balance and Michaelis Menten forms are used to describe substrate level changes as the system equilibrates to a (new) steady state because of nutritional and digestive inputs. Implicit to these models are two basic assumptions: firstly, that *in vivo* metabolic pathways can be represented using the critical transactions modelled from *in vitro* experimental data, and

Table 2. Properties of production and scientific models (Adapted from Boston et al., 2003).

Feature	Production model	Scientific model
Purpose	Predict response	Understand process
Form	Response surface equations	Differential equations (state equations)
Parameters	Polynomial coefficients derived from data fitting	Biochemical reaction properties
Aggregation step	None; model derived from aggregated experiments	Chemical processes aggregated to organ and animal level functions
Solution process	Simple explicit solution of equations	Complex systems of differential equations requiring special software
Outputs	Computed indicators of adequacy of inputs and production cost measures	Steady state solutions to transactions in terms of scientifically significant indicators
Character	Empirical and static	Dynamic and Mechanistic

secondly, that cellular level metabolic processes can be aggregated to the organ level to effectively model whole animal function. Baldwin, at the University of California, and his colleagues (Baldwin *et al.*, 1987a,b,c) have produced a comprehensive integrated model that describes digestion and metabolism of the dairy cow with dynamic, mechanistic equations of physiological processes.

2.2. Production models

Production models are primarily used to portray animal responses to different inputs. They are usually created from collections of response surface models that are developed from animal or herd level experiments. Thus, these models are developed downward. They are valid within the domain of data underpinning the individual response surfaces and are as accurate as the response models themselves.

A theme for the development, refinement and deployment of empirical production models is seen in the development and implementation of the National Research Council dairy cow models (NRC, 1978; 1989; 2001). In 1978, response equations were used to predict crude protein and energy needs of the dairy cow. The 1989 model used a system of protein utilisation that partitioned dietary protein into rumen degradable (DIP) and rumen undegradable (UIP) fractions (NRC, 1985). Growth of micro-organisms in the rumen was driven by energy intake (TDN, NE_L). In 2001, the National Research Council released a new dairy cow model that contains some of the mechanistic approaches in the CNCPS/CPM-Dairy that are described below.

Other empirical production models include the VEM-DVE/OEB (Dutch; Tamminga *et al.*, 1994), AFRC (British; AFRC, 1990; 1992), CSIRO (Australian; CSIRO, 1990) and INRA (French; INRA, 1989) systems. These early production models stimulated more precise thinking and experimentation. Better data were incorporated into newer versions of models. Largely because of concepts in these increasingly precise models, rations for dairy cows usually now contain feed ingredients that are resistant to ruminal degradation. This increases overall efficiency of dairy cow feeding.

The need for more accurate models to define rumen bacterial and whole animal requirements, to assess feed utilisation and to predict production responses led to the development of the Cornell Net Carbohydrate and Protein System (Fox *et al.*, 1992; O'Connor *et al.*, 1993; Russell *et al.*, 1992; Sniffen *et al.*, 1992). The CNCPS is a mix of empirical and mechanistic approaches that describe feed intake, ruminal fermentation of protein and carbohydrate, intestinal digestion and absorption, excretion, heat production, and utilisation of nutrients for maintenance, growth, lactation and pregnancy. When the CNCPS was evaluated with data from individual dairy cows where the appropriate inputs were measured and changes in energy reserves were accounted for, 90% of the variation in actual milk production of individual cows was explained with a 1.3% bias. The model accounted for 76% of the variation in individual cow milk production with an 8% underprediction bias when energy was first limiting and 84% of the variation with a 1.1% overprediction bias when protein was first limiting (Fox *et al.*, 2004).

3. Dairy nutrition software

Dairy nutrition models often do not contain tools for computer assisted ration formulation. Software included with the 1989 and 2001 NRC dairy nutrition models allowed calculation of nutrient requirements and evaluation of rations but did not provide for formulation of rations.

3.1. Auto-balancing

The usual objective of auto-balancing is to produce an 'optimal ration' at the lowest cost. Constraints (minimum and maximum amounts) are set for both nutrients and feed ingredients. Nutritional constraints describe the requirements of cows to perform specific or multiple functions (maintenance, growth, lactation, pregnancy). Nutritional constraints include dry matter intake, energy (metabolisable and net), protein (crude, soluble, bypass and metabolisable), carbohydrates (fibre and non-fibre), fat, minerals; and in the case of newer models like CNCPS/CPMDAIRY, amino acids and rumen available nitrogen (peptides and ammonia) are included. Feed ingredients are selected on the basis of the major nutrients that they provide (i.e. fibre from forages, non-fiber carbohydrates from grains, protein from oilseed meals). Feed constraints are set based on the availability of purchased ingredients and inventory of ingredients on the farm or contracted for purchase. The amount of an ingredient specified is often adjusted by the formulator to take into account a minimum amount that the formulator feels rations should contain or the maximum amount that the formulator feels can be tolerated by the animal. The amount of a feed ingredient should not be limited by high cost because optimisation programs will control the inclusion of expensive feeds. Thus, the auto-balancing (optimisation) task is to find the least cost combination of feed ingredients within their minimum and maximum constraints that provide nutrients that are within the specified minimum and maximum ranges. When the foregoing is achieved, the auto-balancing process has provided a solution to the specifications defined by the formulator.

Ration formulators often are discouraged when the optimisation process does not give a solution as defined above. This simply means that a combination of feed ingredients in amounts within the minimum and maximum ranges cannot provide nutrients within the specified ranges. To find a solution, the formulator should either expand (relax) the feed ingredient and nutrient constraints or include additional ingredients that are good sources of limiting nutrients. Older optimisation methods simply indicated that there was 'no feasible solution'. This provided no direction to obtaining a solution. Newer optimisation methods provide direction by listing nutrient constraints that are not met.

In both empirical models and in CNCPS/CPMDAIRY, nutrients like crude protein, fat, carbohydrates (fibre and non-fibre) and minerals are constant proportions of the ingredient regardless of the amount of feed consumed. Thus, supply of these nutrients is a linear function of intake. In empirical dairy cow models, metabolisable protein and energy (metabolisable and net) values also are not affected by intake and thus are constant. Thus linear programming can be used for auto-balancing. The CNCPS/CPM-DAIRY has a dynamic rumen sub-model wherein the passage rate of feeds (determined mainly by feed intake but also adjusted by ration forage content and particle size) determines the outflow of nutrients from the rumen system. Thus, nutrients like metabolisable protein, metabolisable energy, amino acid content of metabolisable protein, and rumen available protein (peptides and ammonia) are not constant but vary according to feed consumption and ration ingredients. These features of the dynamic digestion models mean that the problem of dairy cow ration optimisation is no longer the province of linear programming and nonlinear programming packages are required. Implementing constrained, nonlinear optimisation is not without problems. If the nutrition model contains discontinuous (break-point) functions, continuous mathematical models must be developed to describe the discontinuous functions (Boston *et al.*, 2000). Whereas a linear programming problem can be solved from any starting point, a nonlinear programming problem requires a 'good' feasible starting point to 'effectively' start the solution process. Finally, a linear programming optimisation problem has just one solution. This is not so for nonlinear optimisation.

Ration formulation was one of the first applications of linear programming. Not only could solutions be found in seconds, but building on contributions of Dantzig (1995) to operational research, we also were able to derive an array of other very helpful economic properties (shadow prices) relating to our optimal solution. For example, we could discover the cost ranges over which feeds within the optimal ration remained there, as well as which amongst the feeds not selected in the optimal ration were candidates for inclusion if cost decreased.

Bath and Bennett (1980) at The University of California were amongst the earliest to employ linear programming to formulate rations for maximum income over feed costs. Galligan and co-workers (1986) at the University of Pennsylvania programmed the 1978 NRC dairy nutrition model into Lotus 1-23 with auto-balancing of rations provided by Einfin. Spartan represented an excellent effort in software development by the group at Michigan State University that was based on NRC models and included auto-balancing (Vandehar *et al.*, 1992).

CPM-Dairy is a combined effort by researchers at Cornell University, the University of Pennsylvania and the W.H. Miner Agricultural Research Institute (Boston *et al.*, 2000). CPM-Dairy contains the CNCPS and is field efficient program for use with growing, lactating and dry cows. The optimisation scheme was developed by the Systems Programming Group at the University of Maryland (Zhou and Tits, 1997) and employs a forward sequential quadratic programming approach. Version 1, released in 1998, also contained a modification of NRC (1989). Versions 2 and 3 only contain the CNCPS and are in beta testing with release anticipated in 2004. Version 3 has expanded carbohydrate fractions, a lipid submodel (Moate *et al.*, 2004) and incorporates NRC (2001) mineral requirements.

3.2. Commercially available software

Table 3 contains a list of commercially available software for formulation of dairy cattle rations. NRC, INRA and CNCPS dairy nutrition models are used in some of the software packages whereas proprietary or user-defined models are used in others. Linear programming is used for auto-balancing in empirical models. In CNCPS, biological values for metabolisable energy, metabolisable protein, passage rate, bacterial yield efficiencies and degradation rate of available fibre, which depend upon feed intake and the ingredients selected, are first estimated and then rations are balanced using linear programming. In CPM-Dairy, a nonlinear optimiser is used to auto-balance rations according to the CNCPS (Zhou and Titts, 1997).

Table 3. Some commercially available dairy nutrition software.

Software	Developers	Dairy nutrition model[1]	Website
Feed ration balancer	Feed management systems	NRC, User defined	http://www.feedsys.com/
CamDairy	Cam Software	Proprietary model	http://epicentre.massey.ac.nz/ Downloads/Software/CD_install.pdf
The consulting nutritionist	Dalex Computer Systems, Inc.	User defined nutrient requirements, CNCPS, NRC	http://www.dalex.com/
CPM-Dairy	Cornell U., U. Pennsylvania, Miner Institute	CNCPS	http://mail.vet.upenn.edu/ ~ejjancze/ cpmbeta3.html
CNCPS	Cornell University	CNCPS	http://www.cncps.cornell.edu/ cncps/ main.htm
Dairy ration system	ACS Computer Services	NRC	http://www.acsdrs.com
Formulate2	Central Valley Nutritional Associates	NRC	http://www.formulate2.com/
INRAtion - PrevAlim	INRA	INRA	http://www.cnerta.educagri. fr/unites/ lpa/lpa.htm#inration
Mixit-Win	Agricultural Software Consultants, Inc.	User defined minimum and maximum nutrient amounts	http://www.asc-mixit.com/
Molly	U. California, Davis	Molly	http://animalscience.ucdavis.edu/ research/molly/default.htm
PCDairy-2	U. California, Davis	NRC	http://animalscience.ucdavis.edu/ extension/pcdairy.htm

Table 3. Continued.

Software	Developers	Dairy nutrition model[1]	Website
RationPro	ProfitSource	NRC, User-defined	http://www.rationpro.com/
RumNut	A.T Chamberlain	AFRC, PDI	http://www.rumnut.com/proginst.pdf
Shield	U. California, Davis	Proprietary model	http://animalscience.ucdavis.edu/extension/shield.htm
SigaDairy	Siga Farm Software	NRC, User-defined	http://www.siga.net
Spartan	Michigan State U.	NRC with modifications	http://www.msu.edu/user/ssl/index.htm
Trilogic	Trilogic Systems	NRC, User-defined requirements, proprietary amino acid/carbohydrate model	http://trilogic-systems.com/

Note: NRC = National Research Council; CNCPS = Cornell Net Carbohydrate and Protein System; INRA = Institut National de la Recherche Agronomique; MOLLY = a dynamic, mechanistic computer model of a dairy cow; AFRC = Agricultural and Food Research Council; PDI (INRA) = Proteines varies reellenment Digestibles dans l'Intestin grele.

4. Application of dairy nutrition models

CPM-Dairy is used by veterinarians, nutrition consultants and the feed industry to evaluate and formulate rations for dairy cattle. At New Bolton Center, our Field Investigation Unit was presented with a case where milk production in a 200-cow dairy was only 30 kg/d although cows were milked three times daily and were treated with bST. Early lactation cows frequently went off feed and had diarrhoea. Faeces contained undigested corn, believed to be high moisture corn.

There were two high production groups (heifers and high cows) and a low production group. The heifers and high production cows were housed in a new free-stall barn with excellent ventilation and cow comfort. Cows were fed three times daily with frequent 'push-up' of feed. Feed bunk space was 1.7 feet per cow with no headlocks, and cows had good water access. Remaining milking cows were housed in a renovated free-stall barn. They were fed twice daily with frequent 'push-up' of feed. Feed bunk space and ventilation were good. Non-lactating cows were housed on a bedded pack.

CPM-Dairy was used to evaluate the existing rations and to formulate new rations. In Table 4 are details of pre-CPM-Dairy and CPM-Dairy rations for the high producing animals (heifers and high cows).

The pre-CPM-Dairy ration was formulated for a target of 39 kg/d milk with 3.6% fat and 3.1% crude protein. According to CPM-Dairy, this ration was low in metabolisable protein and would only support 34.4 kg/d milk. Poor amino acid ratios (Met/ MP=1.89; Lys/MP=6.24) could reduce milk production by 1.6 kg/d so that expected milk on the basis of metabolisable protein and balance of amino acids was only 32.8 kg/d. The ration was marginal in eNDF (21% vs. the guideline of 23% of ration DM) and contained 41.5% non-fibre carbohydrate (NFC). High moisture corn, which has an initial high rate of fermentation in the rumen, contributed a substantial amount of NFC.

Objectives in formulating the new ration were to (1) provide less NFC without compromising total carbohydrate fermentability; (2) increase eNDF; (3) correct the deficiency of metabolisable protein; and (4) improve amino acid balance.

Reducing amounts of high moisture corn (19 vs. 24% DM) and alfalfa silage (15 vs. 27% DM), increasing the amount of corn silage (37 vs. 23% DM) and including soybean hulls (7 vs. 0% DM) reduced NFC (39.0 vs. 41.5%). Although

Table 4. Field application of CPM-Dairy.

	Ration	
	Pre CPM-Dairy	CPM-Dairy
Ingredients (%DM)		
Alfalfa silage	27.14	15.21
Corn silage	22.79	37.16
High moisture corn	24.23	18.95
Soybean hulls		6.92
Soybean meal	5.63	3.91
Corn distiller's grains	5.51	
Dry brewer's grains	3.34	
Whole cotton seed	9.29	10.00
Fish meal		1.38
Blood meal		0.55
Animal-marine protein blend		3.38
Megalac	0.59	
Megalac Plus		0.49
Salt	0.40	0.22
Sodium bicarbonate	0.50	0.80
Mineral and vitamin mix	0.58	1.03
Total dry matter, kg/d	22.46	22.46
Cost, $/day	3.65	3.82
Carbohydrates		
Non fibre[1], %DM	41.5	39.0
Neutral detergent fibre, %DM	30.4	33.3
Effective neutral detergent fibre, %DM	21.0	22.9
Protein		
Crude, % DM	18.6	17.8
Undegraded, % CP	33.7	37.2
Metabolisable, kg/d	2.28[100%][2]	2.49[100%]
Bacterial	1.25[55%]	1.34[54%]
Undegraded	1.03[45%]	1.15[46%]
Metabolisable Amino Acids		
Methionine, g/d	43	55
Met/MP	1.89	2.19
Lysine, g/d	142	173
Lys/MP	6.24	6.94

Table 4. Continued.

	Ration	
	Pre CPM-Dairy	CPM-Dairy
Nutrient limited milk, kg/d		
Metabolisable energy	44.0	43.8
Metabolisable protein	34.4	39.1
Rulquin Ratio[3]	32.8	39.0

[1]Includes sugars, starch, pectin, B-glucans and acids produced during silage fermentation. Silage acids are not fermented further in the rumen and do not provide energy for bacterial growth.

[2]Values in brackets are percentages of metabolisable protein.

[3]Responses to amino acid ratios calculated according to equations in Rulquin and Verite (1993) and added to metabolisable protein-limited milk.

NFC was lower, total fermentable carbohydrates were higher (45 vs. 43% ration DM) because of the high ruminal fermentation of NDF in soybean hulls and corn silage. Increasing ration forage (52 vs. 50% DM) and including soybean hulls increased NDF (33.3 vs. 30.4%) and eNDF (22.9 vs. 21.0%).

Replacing a portion of the soybean meal and all of the corn distiller's grains and dried brewer's grains with animal and marine proteins increased the supply of metabolisable protein from rumen undegraded protein by 0.12 kg (1.15 vs. 1.03 kg/d). Metabolisable protein from ruminal bacteria was increased (1.34 vs. 1.25 kg/d) for two reasons. First, bacterial growth is driven mainly by energy derived from fermentation of carbohydrates. As noted above, the CPM-Dairy ration contained more rumen fermentable carbohydrates (45 vs. 43% DM, 10.05 vs. 9.76 kg/d) than the pre-CPM-Dairy ration. Second, bacterial growth is decreased when rumen pH decreases. The CPM-Dairy ration contained more eNDF than the pre-CPM-Dairy ration. This stimulates cud chewing, increases salivary flow and improves ruminal buffering. More metabolisable protein from rumen escape protein and from bacterial protein alleviated the deficiency of metabolisable protein so that milk yield based on metabolisable protein provided by the CPM-dairy ration was 39.1 kg/d compared to 34.4 kg/d for the pre CPM-dairy ration.

Animal and marine proteins and ruminal bacteria are excellent sources of lysine. Thus, the combination of ruminal escape and bacterial protein improved Lys/ MP (6.94 vs. 6.24). Fish meal, but not blood meal nor rumen bacteria are good sources of methionine. Met/MP (2.19 vs. 1.89) was improved by using Megalac Plus (Alimet in calcium salts of LCFA) to supplement methionine in rumen escape and bacterial protein. Amino acid balance of the CPM-Dairy ration would only reduce milk 0.1 kg/d compared to 1.6 kg/d with the pre CPM-Dairy rations. Thus, on the basis of metabolisable protein supply and amino acid balance, expected milk was 39 kg/d or 6.2 kg/d more than on the pre CPM-dairy ration.

After one week on the new high production rations and a similarly formulated low production ration, total herd milk production was 4 kg/d higher. Cud chewing increased and manure consistency improved. Off-feed problems were dramatically reduced.

Two weeks after the ration change, manure still contained excessive undigested high moisture corn. Grinding the high moisture corn through a 0.635 cm screen reduced corn in faeces and milk production increased another 2 kg/d. Herd milk production now averaged 36 kg/d. For the next 12 months, average milk production for the total herd was 36 to 38 kg/d.

References

Agricultural and Food Research Council, 1990. Technical Committee on Responses to Nutrients, Report no. 5. Nutritive requirements of ruminant animals: energy. *Nutrit. Abstr. and Rev. Series B* **60**:729-804.

Agricultural and Food Research Council, 1992. Technical Committee on Responses to Nutrients, Report no. 5. Nutritive requirements of ruminant animals: protein. *Nutrit. Abstr. and Rev. Series B* **62**:787-835.

Baldwin, R.L., J. France and M. Gill, 1987a. Metabolism of the lactating cow I. Animal elements of a mechanistic model. *J. Dairy Res.* **54**:77-105.

Baldwin, R.L., J.H.M. Thornley and D.E. Beever, 1987b. Metabolism of the lactating cow II. Digestive elements of a mechanist model. *J. Dairy Res.* **54**:107-131.

Baldwin, R.L., J.H.M. Thornley and D.E. Beever, 1987c. Metabolism of the lactating cow III. Properties of mechanistic models suitable for evaluation of energetic relationships and factors involved in the partition of nutrients. *J. Dairy Res.* **54**:133-145.

Bath, D.L. and L.F. Bennett, 1980. Development of a dairy feeding model for maximizing income above feed costs with access by remote computer terminals. *J. Dairy Sci.* **63**:1397-1389.

Boston, R., D. Fox, C. Sniffen, E. Janczewski, R. Munson and W. Chalupa, 2000. The conversion of a scientific model describing dairy cow nutrition and production to an industry tool: The CPM-Dairy project. In: *Modeling Nutrition of Farm Animals* (J.P. McNamara, J. France and D.E. Beever, eds). CAB International, UK, pp. 361-377.

Boston, R., Z. Dou and W. Chalupa, 2003. Models in nutritional management. In: *Encyclopedia of Dairy Sciences* (H. Roginski, J. Fuquay and P. Fox, eds). Elsevier Science, pp. 2378-2389.

Commonwealth Scientific and Industrial Research Organization, 1990. *Feeding Standards for Australian Livestock*. CSIRO Publications, East Melbourne, Australia.

Dantzig, G.B., 1955. A proof of the equivalence of the programming problem and the game problem. In: *Activity Analysis of Production and Allocation* (T.C. Koopmans, ed). John Wiley, NY, pp. 330 335.

Fox, D.G., C.J. Sniffen, J.D. O'Conner, J.B. Russell and P.J. Van Soest, 1992. A net carbohydrate and protein system for evaluating cattle diets. III. Cattle requirements and diet adequacy. *J. Anim. Sci.* **70**:3578-3596.

Fox, D.G. L.O. Tedeschi, T.P. Tylutki, J.B. Russell, M.E. Van Amburgh, L.E. Chase, A.N. Pell and T.R. Overton, 2004. The Cornell Net Carbohydrate and Protein System model for evaluating herd nutrition and nutrient excretion. *Anim. Feed Sci. Tech.* **112**:29-78.

France, J. and J.H.M. Thornley, 1984. *Mathematical Models in Agriculture*. Butterworths. London.

Galligan, D.T., J.D. Ferguson, C.F. Ramberg, Jr. and W. Chalupa, 1986. Dairy ration formulation and evaluation program for microcomputers. *J. Dairy Sci.* **69**:1656-1664.

Institut National de la Recherche Agronomique, 1989. *Ruminant Nutrition. Recommended allowances and feed tables*(J. Jarrige, ed). John Libbey Eurotext, London.

Moate, P.J., W. Chalupa, T.C. Jenkins and R.C. Boston, 2004. A model to describe ruminal metabolism and intestinal absorption of long chain fatty acids. *Anim. Feed Sci. Tech.* **112**:79-105.

National Research Council, 1978. *Nutrient Requirements of Dairy Cattle*. National Academy Press, Washington, D.C.

National Research Council, 1985. *Nitrogen Usage in Ruminants*. National Academy Press, Washington, D.C.

National Research Council, 1989. *Nutrient Requirements of Dairy Cattle, Update 1989.* National Academy Press, Washington, D.C.

National Research Council, 2001. *Nutrient Requirements of Dairy Cattle.* National Academy Press, Washington, D.C.

O'Connor, J.D., C.J. Sniffen, D.G. Fox and W. Chalupa, 1993. A net carbohydrate and protein system for evaluating cattle diets. IV. Predicting amino acid adequacy *J. Anim. Sci.* **71**:1298-1311.

Rulquin, H and R. Verite, 1993. Amino acid nutrition in dairy cows. In: *Recent Advances in Animal Nutrition* (P.C. Garnsworthy and D.J.A. Cole, eds.). Nottingham University Press, UK, pp. 55-77.

Russell, J.B, J.D. O'Connor, D.G. Fox, P.J. Van Soest and C.J. Sniffen, 1992. A net carbohydrate and protein system for evaluating cattle diets. I. Ruminal fermentation. *J. Anim. Sci.* **70**:3551-3561.

Sniffen, C.J., J.D. O'Connor, P.J. Van Soest, D.G. Fox and J.B. Russell, 1992. A net carbohydrate and protein system for evaluating cattle diets. II. Carbohydrate and protein availability. *J. Anim. Sci.* **70**:3562-3577.

Tamminga, S., W.M. Van Straalen, A.P.J. Subnel, R.G.M. Meijer, A. Steg, C.J.G. Wever and M.C. Blok, 1994. The Dutch protein evaluation system: the DVE/OEB system. *Livestock Prod. Sci.* **40**:139.

VandeHar, M., H. Bucholtz, R. Beverly, R. Emery, M. Allen, C. Sniffen and R. Black, 1992. *Spartan Dairy Ration Evaluator/Balancer.* Michigan State Univ., East Lansing.

Zhou, J.L. and A.L. Tits, 1997. *User's Guide for FFSQP Version 3.5.* Electrical Engineering Dept. and Institute for Systems Res. Univ. Maryland, College Park.

Dairy and *Salmonella*

Søren Astrup
Vemb Dyrlæger I/S, Industrivej 32, 7570 Vemb, Denmark

1. Introduction

The reason why *Salmonella* has become so important is partly because it can cause disease and economical loss on farm level, but equally important is the fact that *Salmonella* is a zoonosis. The incidence of Salmonellosis in both humans and animals has increased with the increased intensification of the animal production. A lot of humans become sick and even die from Salmonellosis every year and the bacteria cause considerable loss in the animal production because of decreased production, disease and dead animals.

The most effective point to target the problem is in the primary production which means at farm-level. In cattle the primary *Salmonella* problem is caused by *Salmonella dublin*. In Denmark 25% of the dairy-farms are infected with *Salmonella dublin* and when *Salmonella dublin* acts as a zoonosis is causes severe disease with a high fatality rate.

In a case-study from the author's veterinary practice the addition of mannan oligosaccharides (Bio-Mos®) to the feed were the deciding factor in controlling and eradicating a *Salmonella mrDT104* infection in a large dairy-farm. The author believes that mannan oligosaccharides could prove to be an important tool in the control of the widespread infection of *Salmonella dublin* in the cattle-herds.

2. What characterise *Salmonella* bacteria?

Salmonella are gram negative rodshaped bacteria. They are parasitic in nature and are primarily found in the intestinal canal of animals and humans. More than 2,600 serotypes of *Salmonella* have been isolated so far. Originally the serotypes were named after the disease they caused or after the animal species in which they were first isolated, for example *Salmonella typhi*, *Salmonella enteritidis* or *Salmonella abortusequi*. Later the serotypes were named after the place where they were isolated, for example *Salmonella dublin*, *Salmonella kentucky* or *Salmonella manhattan*.

Some serotypes adhere particularly well to the intestinal mucosal cells in a specific animal species. They are called *Host Specific Salmonella* and they have a tendency to cause septicaemia and chronic carriers (Table 1). The large group of *Non-Host Specific Salmonella* can infect a wide range of animal species. The non-host specific *Salmonella* are normally non invasive and the animals will clear themselves from the infection fairly quickly if they survive (Table 1).

Although animals and especially carriers represent the main harbour for *Salmonella* they survive well in the environment. Temperature and wetness are important for the survival as Salmonella are susceptible to drying and sunlight. *Salmonella typhimurium* can remain viable on pasture and in soil, still water, faeces and slurry for up to 7 months given the right conditions. They will survive well under the warm wet conditions often found in calf- and pig barns but they also survive freezing in water for long periods.

3. Salmonellosis

The disease caused by *Salmonella* bacteria is called Salmonellosis. The disease occurs universally and in all species and is characterised by 1 or more of 3 major syndromes: septicaemia, acute- and chronic enteritis:

- *Septicaemia* ('blood-poisoning' = bacteria in the blood) is the characteristic form seen in young animals in the early stages of an outbreak. The symptoms are profound depression, dullness, high fever and normally death within 24-48 hours.

Table 1. Host specific and non-host specific Salmonellae.

Host specific		Non host specific
Characteristics		
Invasive		Non invasive
Cause septicaemia and gastro-enteritis		Normally only cause gastro-enteritis.
Chronic infection with carrier animals		
Examples		
S. typhi	humans	*S. typhimurium*
S. dublin	cattle	*S. enteritidis*
S. choleraesuis	pigs	*S. senftenberg*
S. gallinarum	hens	...more than 2,000 serotypes

- *Acute enteritis* is the normal symptom found in adult animals during an acute outbreak. There is high fever with severe often bloody diarrhoea. There is severe dehydration and toxaemia, the animal becomes recumbent and dies in 2-5 days.
- *Chronic enteritis* usually but not always succeeds an acute episode. It's characterised by moderate fever, intermittent or persistent diarrhoea with spots of blood and loss of weight.

The mean infective dose is normally 10^5-10^8 bacteria, but in humans it has been shown that a few *Salmonella* bacteria in adipose food like chocolate, cheese or ice cream can cause disease, because the fat will coat and protect the bacteria against the acidic environment in the stomach. It has been shown that as few as 10 *Salmonella typhimurium* bacteria in a piece of cheese have been sufficient to cause disease in humans.

The classic symptoms of Salmonellosis described above goes for all *Salmonella* species and for all animal species, although there are characteristic disease patterns seen in the individual animal species. Salmonellosis in cattle will be described in details below. The disease is always spread by the faecal-oral route and clinically normal carrier animal is a problem in Salmonellosis caused by host specific *Salmonella* bacteria.

4. A zoonosis

Salmonellosis is a zoonosis meaning that it's a disease which can spread from animals to humans. The transmission can be directly from animals to humans, but often the bacteria are transmitted via contaminated food of animal origin for example meat, egg or raw milk.

The main reason for Salmonellosis is contaminated food which is eaten raw or not cooked sufficiently. For example raw eggs in desserts or minced meat and poultry-meat which is not well done. Another common reason is cross contamination between meat and raw vegetables and salad if the same kitchen utensils are used for all cooking steps without proper hygienic measures.

The infective dose of bacteria varies between 10 and 100,000 bacteria per gram food depending on the type of food and the immunological status of the person. A very low dose of bacteria can cause disease in food with a high fat content for example chocolate and snacks. The incubation period is 1-2 days and the symptoms are diarrhoea, stomach pain, fever, headache and vomiting.

The disease normally last a few days to weeks but in 1% of the cases the disease develops to septicaemia with a prolonged and complicated course which in worse case can be lethal.

The best way to prevent Salmonellosis is to combat and preferably eradicate the bacteria in the primary production and to ensure that preventive measures are carried out in all steps from 'stable to table'.

4.1. Prevalence of human Salmonellosis in Denmark

There are on average 2,000 registered cases of human Salmonellosis in Denmark every year (Figure 1). Most likely there are a lot more cases as only a fraction of the cases are microbiologically diagnosed.

After peaking in the late nineties there has been a steady decline in the number of human cases of Salmonellosis which can be attributed to national *Salmonella* control programmes instituted in the poultry- and the pig production.

The three main sources of human Salmonellosis are broilers, eggs and pork (Figure 2). Cattle are only a minor source of human Salmonellosis. Only 40 out of the on average 2,000 human cases of Salmonellosis is caused by *Salmonella dublin* and there are only 0,3-0,5% positives on the routine swabs done on the beef in the slaughterhouses compared to 1,5% in pigmeat and 2,5% in poultrymeat (The Danish Livestock and Meatboard, 2006; Fødevarestyrelsen, 2006; Borck *et al.*, 2003). However the Salmonellosis caused by *Salmonella dublin* is often very severe with sepsis and a complicated course.

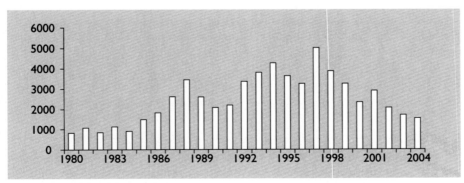

Figure 1. Number of human cases of Salmonellosis in Denmark 1980-2004 (www.mave-tarm.dk).

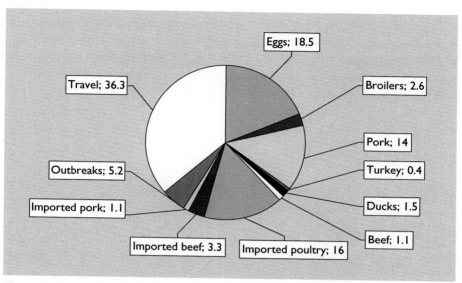

Figure 2. Distribution of sources of human Salmonellosis in percent of total (Borck et al., 2003).

5. *Salmonella* in cattle

In 2003 *Salmonella* was isolated in 74 cattle herds in Denmark (Table 2). The predominant serotypes were *Salmonella dublin* (54%) and *Salmonella typhimurium spp.* (44.6%)

Table 2. Salmonella *serotypes isolated in Danish cattle herds in 2003 (Borck et al., 2003).*

Serotype	Number isolated	Percentage of total
Salmonella dublin	40	54.0 %
Salmonella infantis	1	1.4%
Salmonella typhimurium	32	43.2%
Salmonella typhimurium MRDT104	1	1.4%
Total herds examined	74	100%

Although all *Salmonella* species can cause disease in all ages of cattle the tendency is that *Salmonella dublin* causes clinical disease in calves while *Salmonella typhimurium* causes disease in adult cattle. Among the *Salmonella typhimurium* serotypes the multiresistant DT104 has caused special concern because of the difficulties treating the infection if the bacteria spread to humans.

5.1. Salmonella dublin

5.1.1. Pathogenesis

Salmonella dublin is specially adapted to infecting cattle. They are most often isolated in cattle although they can infect other species. *Salmonella dublin* enters the organism through the mouth or the nose via contaminated feed or water or by licking on other animals, walls or bedding contaminated with faeces (Figure 3). The bacteria pass through the rumen and abomasum where a large proportion is killed before they enter the small intestine. The *Salmonella* bacteria multiply

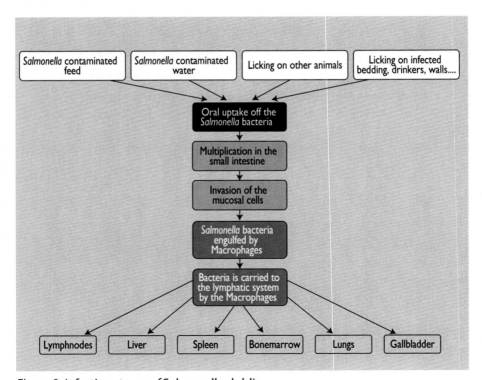

Figure 3. Infection stages of Salmonella dublin.

in the small intestine and start to invade the intestinal wall, where they are attacked by macrophages from the non-specific immune response. Normally bacteria engulfed by Macrophages will be killed, but *Salmonella dublin* is able to survive inside the macrophages. Inside the macrophages they are passively transported via the lymph and blood stream to the lymph nodes and organs like liver, gallbladder, bone marrow and lungs.

Depending on the immunological status of the host 5 scenarios can be seen (Figure 4):
1. The infection can be stopped by the immune system and the animal cleans itself from the infection.
2. Acute clinical case with diarrhoea, fever, pneumonia and intoxication.
3. *Latent carrier* with the infection persisting in the lymph nodes but no bacteria in the faeces.
4. *Active carrier* with no clinical symptoms but constantly shedding bacteria in the faeces.
5. *Passive carrier* where the bacteria passively passes through the animal without affecting the animal.

Generally clinical disease is primarily seen in young animals while adult animals develop into latent and active carriers.

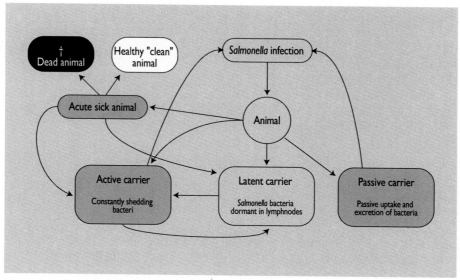

Figure 4. Infection scenarios of Salmonella dublin hosts.

5.1.2. Transmission of the infection within the herd

Within the herd the latent carrier plays a key role in the infection dynamic (Figure 4). Often the latent carrier is an adult cow. Around calving the cow is subject to a lot of stress and if she is a latent carrier she often becomes an active shedder (Figure 5). The newborn calf is very susceptible to infection and thus there is a great risk of transmission of infection from the cow to the newborn calf. When moved to the calf stable there is a great risk of transmission of infection between calves. The next hazard is when the calves are moved to deep litter pens. They are stressed and latent carriers become active thus infecting the other calves (Figure 5).

Heifers above one year holds a great risk of becoming latent carriers when infected and the same goes for cows infected close to calving. Pregnant cows will often abort when infected for the first time.

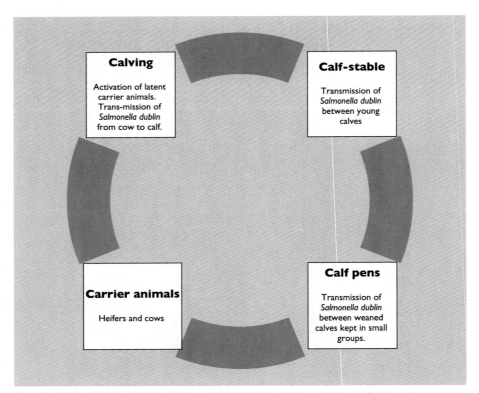

Figure 5. Infection dynamic of Salmonella dublin *during cattle production.*

As immunity in the herd builds up, the clinical disease might disappear for months to a few years only to flare up suddenly when immunity is suppressed or not present in an age group or a section of the stable.

5.1.3. Transmission of the infection between herds

The main cause of transmission of infection between herds is trade, but the infection can also be transmitted via cattle shows or when two herds share a pasture. Passive transmission can occur by service personnel (vets, A.I. technicians, hoof trimmers) and if two herds share machinery.

5.1.4. The Danish National Surveillance Program

Around 25% of the Danish dairy-farms are infected with *Salmonella dublin*. The infection causes considerable economical loss on the individual farm and possesses a zoonosis risk to the consumers. In 2002 a National Surveillance Programme was set up for *Salmonella dublin*. Based on serological testing of blood and milk samples cattle herds are divided into three groups:
- Level 1: The herd is most likely free from *Salmonella dublin*.
- Level 2: *Salmonella dublin* is most likely present in the herd or the herd has an unknown status.
- Level 3: *Salmonella dublin* has been isolated from the herd or the farmer has purchased animals from a Level 3 farm.

The *Salmonella Dublin* Surveillance Programme is linked to the BVD eradication programme. Milk and blood-samples collected as a part of the BVD-programme will also be tested for *Salmonella* antibodies. Every three months dairy-herds are tested on a milk sample collected by the dairy and beef-herds are tested on three blood samples collected in the slaughterhouse or by blood sampling three animals in the stable. Based on the level of antibodies in the milk- and blood-samples the herd are placed in one of the three levels. For herds with less than 25 animals the test will be valid for 12 months and for herds with more than 25 animals the test will be valid for 4 months. The distribution for March 2005 is shown in Table 3.

The overall objective is to decrease the spread of *Salmonella dublin* between herds by setting up a system which encourages trade between negative herds. The levels are published on the internet and the farmer has an obligation to inform trade partners about his *Salmonella*-level.

Table 3. Surveillance level distribution of Salmonella dublin *in March 2005 (Danish Dairy Board, 2005).*

Salmonella dublin level	Percentage of farms in each level	
	Dairy farms	Beef farms
Level 1	82.2%	50.5%
Level 2	17.7%	49.4%
Level 3	0.1%	<0.1%
Total	100%	100%

Level 3 farms will be placed under official restrictions and supervision and the animals are slaughtered under special hygienic conditions. It's voluntarily whenever a level 2 farm wants to set up an eradication program to free the farm from the infection.

5.1.5. Eradication program on herd level

It's difficult to eradicate the infection and it will cost a lot of effort and money, but in the long run it will save the farm from a lot of problems and production losses. The eradication scheme will be tailored to the individual farm, but it will consist of the following components:
- Carrier animals can be identified by blood-testing all cows and heifers over 12 months 3 times yearly – animals with a constant high positive reaction are carrier animals and should be culled.
- Calving cubicles must be clean with fresh new bedding.
- The newborn calf must be moved to a clean separate box immediately after calving to minimise the risk of transmission of salmonella from the cow.
- The calf should be given 20 ml *Salmonella dublin* serum on day 1 and day 17.
- The calves should be kept in a single-box system until 12 weeks.
- After 12 weeks the calves moved from the single-box system should be kept in small groups.
- All in – all out.
- Keep a one-way flow through the herd of animals, personnel and tools.
- Plenty of light, fresh air and a dry environment.

In the author's opinion the addition of mannan oligosaccharides like Bio-Mos® to ration of the dry cows, the newly calved cows and the calves until 6 months could help reducing to transmission of *Salmonella* bacteria.

5.2. Salmonella typhimurium

Salmonellosis in cattle caused by *Salmonella typhimurium* occurs sporadically and even though it can be highly fatal to the individual animal and to some extend to the individual farm it is not really a serious problem.

The source of infection is rodents, birds and feedstuff and the highest incidence of the disease is seen in the summer when the cattle are in close contact with the reservoir hosts (Figure 6).

The disease primarily affects the adult animals and the symptoms are sudden anorexia and a sharp fall in the milk yield, followed by severe diarrhoea with blood and mucus and high fever. Without early treatment 75% of affected animals will die within 2-5 days. Pregnant animals will often have an abortion.

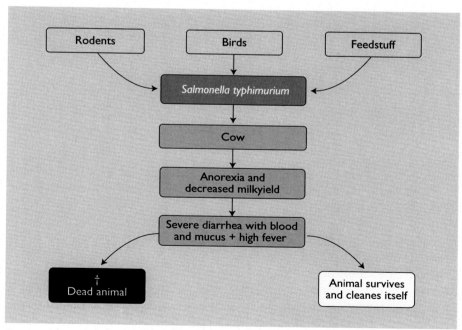

Figure 6. Infection stages of Salmonella typhimurium.

If the source of the bacteria is the feed, the symptoms will start 2-3 weeks after the introduction of the contaminated feed. The number of *Salmonella* in the feed is normally low and it takes a few weeks for the bacteria to multiply to a sufficient number to cause disease. A large proportion of the animals feed with the contaminated feed will show symptoms similar to those described above for the Salmonella transmitted by rodents and birds.

Herds with acute outbreaks of Salmonellosis will be placed under official restrictions and supervision for at least three months after the last symptoms on Salmonellosis. Normally the disease disappears after a few months, but in rare cases the infection can persist in the herd for long periods.

Treatment and control consist of the following components:
- A faeces-sample should be send to a laboratory for diagnosis and testing of the antibiotic resistance pattern.
- Treat sick animals with broad-spectrum antibiotics, fluid therapy, astringent preparations and painkillers.
- Isolate sick animals.
- Restrict the movement of animals around the farm.
- Optimise hygienic precautions.
- One-way flow of personnel and tools through the farm. Start working with the youngest animals and end with the cows.
- Inform the personnel about the zoonosis risk.

5.3. Salmonella multiresistant DT104

5.3.1. Characteristics of Salmonella mrDT104

Multiresistant *Salmonella DT104* (*mrDT104*) is a *Salmonella typhimurium* subspecies characterised by the phagetype DT (distinguished type) 104, a single 60 megadalton plasmid and resistance to at least the five following antibiotics: ampicillin, chloramphenicol, streptomycin, sulfonamides and tetracyclines.

The genes coding for this antibiotic resistance are characterised by being chromosomally integrated as opposed to the normal plasmid-mediated resistance. Chromosomal resistance is a mechanism by which bacteria can retain resistance patterns permanently, even in the absence of the selective pressure of antibiotics. Therefore withdrawal of antibiotics is unlikely to have any impact on the resistance pattern.

The genes coding for the multiresistance is a part of a 'cassette' (integron). A cassette can easily be exchanged with other totally unrelated bacteria. Apart from the chromosomal resistance *Salmonella mrDT104* easily takes up plasmid-mediated resistance and those become resistant to more than the 5 antibiotics mentioned above. Especially an increasing resistance to flouroquinolones has caused concern and has lead to the ban of use of flouroquinolones in Danish animal husbandry.

5.3.2. Why are we specially concerned about mrDT104?

One of the major concerns regarding *mrDT104* is that it is associated with higher hospitalisation and mortality rates in humans than other *Salmonella* species. In addition to this come the difficulties treating the infection because of the multi-resistance. A UK study showed than 41% of the human cases were hospitalised and 3% of the patients died (Dargatz *et al.*, 1998).

Children have shown to be especially susceptible to infection with *mrDT104*. The symptoms in humans are similar to other *Salmonella* infections: diarrhoea, fever, headache, nausea, bloody stool and vomiting.

5.3.3. Symptoms in an outbreak on a cattle-farm

The symptoms of Salmonellosis caused by *mrDT104* are similar to any other *Salmonella typhimurium* outbreak: sudden anorexia and a sharp fall in the milk-yield, followed by severe diarrhoea with blood and mucus and high fever. The disease primarily affects the adult cows, but off course calves can be affected to.

The clinical course in the herd is highly variable: in some herds a large proportion of the cows and calves are affected while in other herds only a few postpartum cases or a few cases in the calves are observed even though the environment is heavily contaminated. Probably differences in management and the farms environment account for the different clinical pictures that are seen on different farms.

5.3.4. Reservoirs of mrDT104 during an outbreak

Reservoirs for *mrDT104* are not well established and the bacteria has been isolated from a wide variety of animal species including cattle, sheep, goats, pigs, birds, dogs, cats, mice, horses and even flies. This wide range of reservoirs makes control and eradication even more difficult.

Clinical normal carrier cows have been shown to shed the pathogen for up to eighteen months following a clinical outbreak. One study identified a clinical normal carrier cow shedding more than 1 million bacteria per gram faeces for more than 6 months (Gay, 2004).

Rodents can be a significant reservoir during an outbreak. A single rodent faecal pellet has been shown to contain 10,000 *Salmonella* bacteria, thus 100 rodent faecal pellets can be enough to cause a clinical case in an adult dairy-cow (Gay *et al.*, 1999).

Clinically normal dogs and cats have also shown to be a significant reservoir of bacteria during an outbreak. Thus the farm dog or cat can transmit the disease between the cattle but equally important is the risk that the pets will transmit the disease to the household where especially the children will be at risk.

It has been shown that flies can act both as mechanical and biological vector as the salmonella bacteria are able to multiply in the flies under the right conditions. In several case studies *mrDT104* positive flies has been found in all fly-collections (Gay *et al.*, 1999).

5.3.5. Danish legal regulations in pig- and cattle farms

In Danish pig herds *Salmonella mrDT104* is treated in the same way as other *Salmonella* infections. Thus there is no Zoonosis Supervision in pigs, but the farmer is obliged to set up a plan to reduce the *Salmonella* load in the herd and there are certain rules concerning transport to the slaughterhouse, trade and management of slurry. There is still a zero-tolerance of *mrDT104* in the pork-meat.

Since the occurrence of *mrDT104* in cattle is very low in Denmark (Figure 7), the cattle industry wished to maintain a zero-tolerance of *mrDT104* in Danish cattle to reduce the risk of spreading the infection between herds and to humans.

When *Salmonella DT104* is detected on a cattle-farm the following things happens:

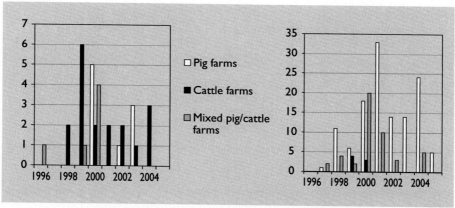

Figure 7. (a) Number of farms with clinical outbreak of mrDT104 *during 1996-2004. (b) Number of farms with restrictions because of* mrDT104 *during 1996-2004 (Fødevarestyrelsen, 2006).*

- The farm will be placed under official restrictions and zoonosis supervision.
- An action plan must be set up within a month to eradicate the infection. The plan must be set up in co-operation with the veterinarian and other relevant advisors – the plan must be approved by the Government Vet.
- Animals can not be sold to other farms.
- Animals can only be sent to the slaughterhouse on special days where they will be slaughtered under special hygienic precautions and the meat will be hot water treated.
- There are special rules on how to handle the manure.
- Service-personnel like veterinarians and AI-technicians must visit the farm as the last farm-visit that day and should wear special boots and coveralls that are left on the farm.
- The milk is collected as the last farm on the route.

To set the farm free from the official restrictions and the zoonosis supervision it must test negative for *Salmonella mrDT104* in two consecutive free testing rounds done with a minimum interval of 30 days. Other farms who have had contact to the infected farm need to be identified and examined to make sure that the infection has not spread to other farms.

The free-testing consist of a number of faeces-samples collected from the rectum or picked from the floor. The samples are pooled 5 and 5 at the laboratory and tested for *Salmonella*. *Salmonella* positive samples are phage-typed and their antibiotic resistance pattern is determined to see if it is a *mrDT104* or just an 'ordinary' *Salmonella*. The number of faeces-samples that needs to be collected in each free-testing round should ensure that the prevalence with 99% certainty is below 5% if all the samples are negative. is given in Table 4. The Government veterinarians can decide that additional samples should be collected from the slurry.

Table 4. The number of faeces samples to be collected for each farm size (Anonymous, 2005).

No. of cattle in the stable	No. of faeces-samples to be collected
<60	All animals should be tested
60-99	60 faeces-samples
100-199	75 faeces-samples
>200	100 faeces-samples

6. *Salmonella DT104* multiresistant on a dairy-farm: a case from practice

6.1. Presentation of the farm

The case concerns a well managed dairy-farm with 160 cows plus the young stock. Furthermore they have a porker-production divided on two farms. The milk-yield is medium high with 8,500 kg milk with 4% fat per cow per year and the cell-count is between 200,000 and 300,000. The cows are kept on slatted floor with free stalls with rubber mattresses. The feeding is TMR (Total Mixed Ration) based on maize-silage.

The cows are divided into 5 pens:
1. high-yielding;
2. mid-lactation;
3. pregnant cows;
4. dry-cows;
5. deep-litter section with cows with leg-problems and newly calved-cows.

The calves are kept in individual cubicles in two sections which allow for an adapted all in – all out system. After weaning they are transferred to deep-litter pens with 5 calves and latter a new pen with up to 15 calves. The young heifers are kept on slatted floors on the main farm until they are ready for breeding. When ready for breeding the heifers are moved to a second farm where a bull will service the 75 heifers that are constantly here to be breed.

The bull-calves are kept on deep-litter until slaughter. Some of the pigs are kept in a stable in the same building as the heifer-farm.

6.2. Break-down of the development of the disease

In May 2000 *Salmonella mrDT104* was detected in the pig-farm and three months later in August the first clinical case of Salmonellosis caused by *Salmonella mrDT104* was diagnosed in a dairy-cow.

From August 2000 until January 2001 the infection spread to the whole farm with clinical disease in all age groups. After January 2001 the number of clinical cases decreased, but regular surveillance faeces-sampling constantly showed positive samples for *Salmonella mrDT104* in every test round despite an intensive eradication programme where a lot of management procedures were adjusted (Figure 5).

On 29th April 2004 a mannan oligosaccharide (Bio-Mos®) was added to the diet of all animals on the farm. Monthly surveillance faeces-samples were taken in all age-groups on the farm and within a month the number of positive samples started to decline and on 30th November 2004 the surveillance faeces samples tested negative for the first time in more than four years (Figure 8).

Equally interesting was the change in the resistance pattern of the *Salmonella DT104* after the addition of Bio-Mos® to the feed. Within a few months the resistance-pattern changed from multiresistant *DT104* to non-multiresistant *DT104* (Figure 9). The resistance was monitored by testing the bacteria's sensitivity towards chloramphenicol. Resistance towards chloramphenicol was interpreted as multiresistance as chloramphenicol is one of the five antibiotics towards *Salmonella mrDT104* always is resistant.

Bio-Mos® directly binds pathogens in the intestinal canal thus preventing the bacteria from multiplying. But Bio-Mos® will also inhibit the transfer of antibiotic-resistance between bacteria by binding the plasmids or the 'cassettes' carrying the genes coding for antibiotic resistance.

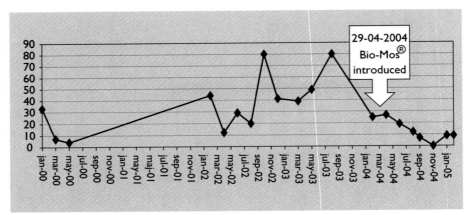

Figure 8. Number of positive Salmonella mrDT104 *faeces samples taken from the dairy farm during 2000-2005.*

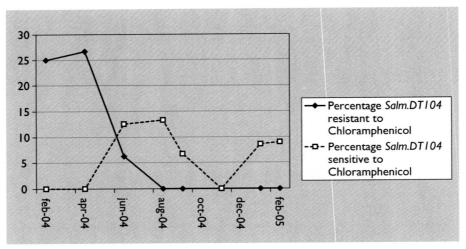

Figure 9. Resistance and sensitivity of Salmonella DT104 *strains taken from faeces samples to chloramphenicol after the addition of Bio-Mos®.*

After the finding of *Salmonella*-free surveillance faeces samples taken on 30th November 2004 it was decided to start the two free-testing rounds (Figure 5). 160 faeces-samples were taken on 5th January 2005 and another 160 samples on 8th February 2005. The samples were pooled five and five and tested for *Salmonella*. Three samples tested positive for *Salmonella DT104* on samples from the 5th January and six from 8th February. Fortunately they were all of the non-multiresistant type and on the 28th February 2005 the farm was set free from the official restrictions and zoonosis supervision after being infected for 4½ year.

6.3. Conclusion

Apparently the addition of the mannan oligosaccharide Bio-Mos® to the feed was the deciding factor in controlling and eradicating the *Salmonella mrDT104* infection on the dairy farm.

References

Anonymous, 2005. Bekendtgørelse om Salmonella hos kvæg og svin mv., BEK nr. 112 af 24/02/2005.

Borck, B., T. Hald, P.C. Sørensen and S. Ethelberg, 2003. *Annual Report on Zoonoses in Denmark*. Ministry of Food, Agriculture and Fisheries, p. 3-18.

Danish Dairy Board, 2005. *Oversigt over Salmonella Dublin-niveauer*. Dansk Kvæg, Mejeriforeningen.

Dargatz, D.A., S.J. Wells, P.J. Fedorka-Cray and J. Akkina, 1998. The Veterinarians Role in Diagnosis, Treatment and Prevention of Multidrug Resistant Salmonella typhimurium DT104. *The Bovine Practitioner*, **No. 32.**2:1-6.

Fødevarestyrelsen (Danish veterinary and food administration), 2006. Ministry of Family and Consumer Affairs. http://www.foedevarestyrelsen.dk

Gay, J., 2004. Bovine Herd Salmonellosis, version 2, updated 2004. Washington State University. Available from: http://www.vetmed.wsu.edu/courses-jmgay/FDIUHerdSalmonella.htm

Gay, J.M., C. Gay, T. Besser, M. Davis, D. Hancock, L. Pritchett and D. Rice, 1999. Salmonella DT104 and Dairy Farms: Lessons from an Emerging Pathogen. Dairy Farm Food Safety and Quality Assurance Symposium, Burlington, Vermont, 1999. Available from: http://www.vetmed.wsu.edu/courses-jmgay/FDIUSalmonellaOverview.htm

Tarminfektioner Monitor (Gastroenteritis monitor Denmark), 2006. http://www.mave-tarm.dk and http://www.germ.dk

The Danish Livestock and Meat Board, 2006. http://www.meatboard.dk

The Selenium dilemma: what is the role for EU approved Sel-Plex®?

Steven A. Elliott PAS
Alltech, Inc.

1. Introduction

Selenium, like the other trace minerals is necessary to sustain life and is essential for basic physiological functions in cattle. While the daily requirement for trace minerals is obviously small, their importance to and impact on production livestock are well documented in research. Fortunately, the difference between deficiency and toxicity with most of these trace minerals is fairly broad. This allows for a wide range of supplemental feeding recommendations and practices to be implemented without causing too many problems. Selenium, on the other hand, was recognised as a potentially toxic mineral long before it was identified as an essential nutrient. This is one of the main reasons why regulations are in place in many parts of the world to ensure that the industry is cognizant of the governments concerns.

It wasn't until 1957 that selenium was even recognised as being a required dietary trace mineral. That was when a German biochemist demonstrated that liver necrosis in rats fed torula yeast could in fact be prevented by simply supplementing the yeast that they were eating with selenium (Schwarz and Foltz, 1957). This was just the beginning of a long battle to get selenium approved as a feed additive because of the narrow range between deficiency and toxicity, coupled with the concern that selenium was an environmental toxicant. These concerns lead to an additional 30 years of research and political debate before finally in 1987 selenium was FDA approved for supplementation at 0.3 ppm in complete feeds for the major food-producing animals.

2. Selenium concentrations in soil and feed ingredients

The concentration of Se in the soil and its availability to crops vary greatly from one geographic region to another. Selenium in soil is present largely in inorganic forms. Selenium is not a requirement of plant growth but, depending on plant species and soil conditions (i.e. pH, moisture and aeration) plants will uptake selenium (Jacques, 2002). Improved agricultural practices such as better use of fertiliser, soil aeration, and improved pasture species are negatively correlated with plant selenium concentration (Gupta and Watkinson, 1985). As such, areas which were previously selenium adequate have now become marginally selenium deficient (Oldfield, 2002). To improve this situation, several areas around the world apply selenium prills in fertilisers. Pasture selenium levels tend to peak after application of these prills and then tail off. The selenium level in these pasture forages can vary substantially throughout the year. Therefore, the rates used are conservative to avoid potential toxicity from extra uptake after initial application. So, selenium fertilisers have only ever gone part way towards achieving long term fully adequate selenium intakes, especially for high producing dairy herds.

3. Selenium supplementation in the international context

Supplementary feeding levels of selenium vary from country to country (Figure 1). Grace (1994) recommends 0.03 ppm calculated by using a factorial model based on the population of animals which had never had high selenium supplementation. It is well established that increasing dietary selenium increases selenium concentrations in the blood, serum, tissues and milk of the cow (NRC, 2001; House and Bell, 1994; Conrad and Moxon, 1979; Grace *et al.*, 1997). As such, the factorial model approach, which is based on the deposition of selenium in body tissues, conceptus, and milk (Grace, 1994; NRC, 2001), will inevitably underestimate the requirements for dairy cattle.

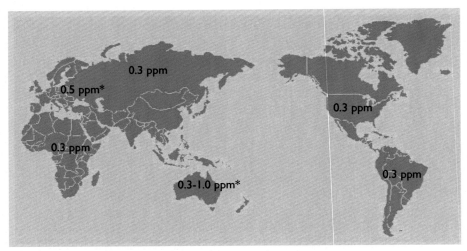

Figure 1. Regional restrictions for selenium supplementation.
* Denotes total diet limitations and/or special considerations for Sel-Plex®.

The NRC recommends that selenium supplementation should be 0.3 ppm. NRC (2001) recommendations are based on extensive observations and experimentation presented to the US Food and Drug Administration which established the links between selenium status, animal health and production responses (Ullrey, 1992). Similar selenium levels are currently targeted by a significant number of dairy vets and nutritionists around the world based on observed positive responses in sub-clinical deficiency syndromes such as retained placenta, reduced fertility, and impaired mammary health (Table 1).

Data from both Maus *et al.* (1980) and Weiss *et al.*, (1990) indicate that plasma selenium concentrations plateau at 1,016–1,397 nmol/L (or 0.08-0.11 mcg/ml) which corresponds to dietary inclusion levels of 0.25–0.3 ppm (Figure 2). In another study, Jukola *et al.* (1996) recommended that in order to optimise the effects of selenium on mastitis, concentrations of selenium should be >0.18 (mcg/ml) in whole blood or ~0.08 (mcg/ml) in plasma. This also corresponds with the recommendations of Maus *et al.* (1980) and Weiss *et al.* (1990). As such, it is suggested that this relationship may be indicative of optimal selenium status in cows.

Original recommendations in Europe were 0.1 ppm. However due to better understanding of the requirement for selenium in dairy cattle diets, much higher levels, similar to NRC (2001) are now typically used there and the maximum tolerable level has been raised to 0.50 ppm (Pehrson, 1993).

Table 1. Recommended feeding levels of selenium for dairy cattle around the world.

Country	Reference	Recommendation ppm
Australia/New Zealand	Macky, 1991	0.30
North America	NRC, 2001	0.30
European Common Market	Pehrson, 1993	0.50

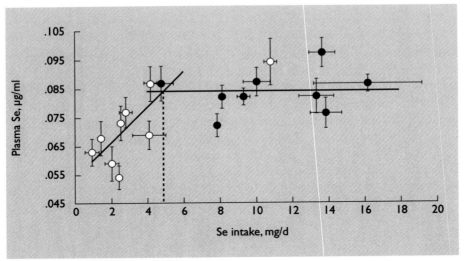

Figure 2. Relationship between selenium intake and plasma concentrations of selenium in dry (O) and early lactation cows (●). Points are herd averages. Adapted from Weiss et al. (1990).

4. Supplemental selenium

Selenium supplements for dairy cows have been traditionally based on inorganic selenium (sodium selenate, barium selenate, sodium selenite). More recently organic selenium supplements have been introduced, and only one, Sel-Plex® selenium yeast, has been scientifically proven and reviewed by the various regulatory commissions around the world (i.e. FDA, EU, and APVMA).

Sel-Plex selenium yeast is manufactured by growing yeast in a nutrient medium low in sulphur and rich in selenium. The yeast forms a high proportion of selenoamino acids using selenium in place of sulphur, creating SeMet, SeCys and other selenoamino acids, the forms of selenium which occur naturally in plants (Jacques, 2002). As reviewed by Weiss (2003) several studies evaluating the impact of different forms of supplement on whole blood selenium and milk selenium have shown that organic selenium in the form of Sel-Plex is more highly available to dairy animals than are inorganic forms.

5. Toxicity

It cannot be forgotten that selenium potentially functions as a toxin as well as an essential element. However, recent ruminant research indicates that toxicity levels are much higher than once thought.

NRC (2001) indicates that the maximum tolerable level for all species is 2 ppm. This estimate was based on data from monogastric animals and does not consider the fact that ruminants have less efficient absorption due to reduction of inorganic selenium sources in the rumen (Wright and Bell, 1966). The NRC (2001) indicates that chronic toxicity (alkali disease) is caused by the ingestion of 5 – 40 ppm of selenium for several weeks or months. More recent work (Cristaldi *et al.*, 2005; Davis *et al.*, 2006a,b) evaluating both inorganic selenium and selenium yeast in sheep diets found that concentrations greater than 10 ppm of dietary selenium elicited no signs of selenotoxicosis.

6. Selenium and its impact on production

Even though improvements in growth or milk production are rare, supplementing selenium to Se-deficient diets generally elicits a positive response in animal health (Weiss, 2003). Additionally, research with dairy and beef cattle has found that selenium supplementation to Se-deficient diets and/or Se-deficient areas can result in:
1. a reduction in milk somatic cell counts (Weiss *et al.*, 1990);
2. a lower incidence and severity of clinical mastitis (Smith *et al.*, 1997);
3. fewer retained placental membranes (Harrison and Conrad, 1984); and
4. improved reproductive parameters (Arechiga *et al.*, 1994).

Each of these areas can have an important impact on the profitability of a dairy farm and by simply improving the selenium status of the animals, the economic value can be appreciated.

7. Selenium and mastitis

The role of selenium in mammary health is one of the best understood response mechanisms. Selenium plays a functional role in the antioxidant GSH-Px which protects cells and body tissues from the auto-oxidative damage of Reactive Oxygen Metabolites (ROM) by leukocytes during phagocytosis and pathogen kill activity. Infusion of a massive amount of neutrophils into the mammary gland is the primary defence apparatus to combat infection.

Selenium status of lactating dairy cows is related to the incidence of mastitis and high somatic cell counts (SCC). A survey of Ohio dairy farms found a strong relationship between herd selenium and SCC (Figure 3). The herds with higher blood selenium concentrations had lower bulk tank SCC than herds with lower blood selenium levels (Weiss *et al.*, 1990). The range of somatic cell counts in these farms was between 400,000 and 100,000, making them well within normal farm averages within the Ohio dairy region.

This reduction in somatic cell count not only improves milk quality, but also can lead to additional milk production over the total lactation of a cow (Table 2). The chart shows how researchers have calculated that a 50% reduction in bulk tank somatic cell count could result in up to 400 lbs. of additional milk/cow/year in second lactation animals.

Whole blood selenium is considered the best indicator of selenium status due to the incorporation of Se into developing red blood cells. A selenium level ranging between 130-150 ng/ml has historically been considered to be the adequate range. Some research suggests that higher blood selenium levels are beneficial for addressing specific mastitis pathogens (Jukola *et al.*, 1996). It was determined that cows with blood selenium of 200 ng/ml or greater had 18% fewer mastitis infections than cows whose blood selenium levels were 150 ng/ml or less (Figure 4).

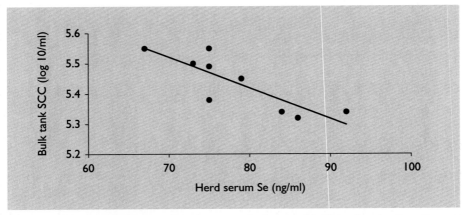

Figure 3. Relationship between herd serum selenium and bulk tank somatic cell count in Ohio dairy herds (Weiss et al., 1990).

Table 2. Relationship between somatic cell counts and lost milk production during the first and second lactation.

Average SCC Score	Lactation average SCC (cells/ml)	Difference in milk yield Lactation 1 (lbs/305 days)	Lactation 2 (lbs/305 days)
2	50,000	-	-
3	100,000	-200	-400
4	200,000	-400	-800
5	400,000	-600	-1,200
6	800,000	-800	-1,600

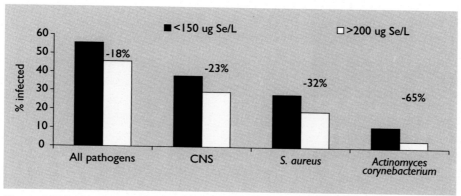

Figure 4. Prevalence of infections of cows with high and low blood selenium levels.

8. Selenium and reproduction

Selenium has been identified as one of the many factors which influence reproductive parameters in dairy cows. The mechanism of selenium deficiency and impaired reproductive function is not well understood. It is postulated that selenium could have an impact through its influence on immune function (Spallholz *et al.*, 1990), uterine contractility (Segerson *et al.*, 1980), thyroid hormone metabolism (Arthur *et al.*, 1993; Wichtel *et al.*, 1996) or synthesis of prostaglandins (Hemler *et al.*, 1979; Hemler and Lands, 1980; Marshall *et al.*, 1987; Reddanna *et al.*, 1989). More recently Arnér and Homgren (2000) cited that in mammals, intra and extra cellular thioredoxin synthesis may be involved in establishment of pregnancy.

Harrison and Conrad (1984) observed a decrease in days to first service and days open in selenium supplemented cows. Similar responses were found by McClure *et al.* (1986), and Arechiga *et al.* (1994). The Arechiga trial showed that supplementing dairy cows with selenium improved the services per conception, pregnancy rates at first service, and resulted in less days to conception (Table 3).

Further improvements may be appreciated with the inclusion of organic selenium, i.e. Sel-Plex, in the diet (Figure 5). In a large trial conducted in the southern US, replacing the inorganic selenium (sodium selenite) with Sel-Plex® significantly improved second service conception rates (Silvestre *et al.*, 2006).

Table 3. Effect of selenium on post-partum reproduction in dairy cows.

	Control	Se	P
Retained placentas	10/99	3/99	0.06
Pregnancy at 1st service	25%	41%	0.02
Services/conception	2.8	2.3	0.03
Days to conception	141	121	0.06

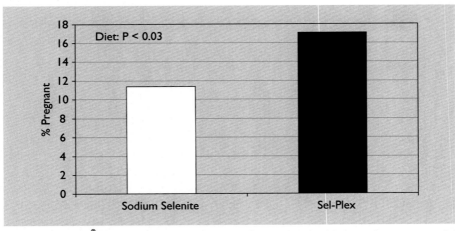

Figure 5. Sel-Plex® significantly improved second service conception rate in a commercial dairy herd.

9. Meeting the selenium requirement of dairy cattle

Continuing problems with dairy cows, like the mastitis and reproductive disorders mentioned above, suggests that current practices of selenium supplementation using sodium selenite may not be adequate. Scientific recommendations, like the NRC, are generally based on providing adequate trace mineral levels to prevent measurable deficiencies, not on adequate levels to optimise animal health.

Another reason why the above mentioned selenium limits and/or sources may need to be reconsidered is that milk production per cow has increased by 16% in the last ten years and it stands to reason that higher-producing modern dairy cows require more selenium than their lower producing ancestors. Finally, it is well documented that the currently available inorganic selenium sources, sodium selenite and sodium selenite, are poorly absorbed and utilised by ruminants.

Inorganic selenium (selenite) is absorbed much less efficiently by ruminants than monogastrics (Gerloff, 1992). Absorption of selenite by ruminants has been reported at 29% (Van Saun, 1990) and between 17 and 50% (Harrison and Conrad, 1984). Poor absorption of inorganic selenium is likely due to the ruminal environment where oxidised selenite or selenate is in large part reduced by ruminal microbes to insoluble and unavailable elemental selenium which is excreted via the faeces (Van Saun, 1990). Other dietary factors also influence the availability of inorganic selenium. Dietary concentrates alter the ruminal reduction capacity with high concentrate/low pH presumably increasing the amount of inorganic selenium that the microbes would make unavailable to the animal (Gerloff, 1992). This variation in rumen acidity may be one reason why the response to a given amount of selenite can vary from farm to farm in the same region. In addition, other minerals including sulfur and iron interfere with selenium absorption (Gerloff, 1992).

10. Improving the selenium status of dairy cattle

Given the legal limits on supplemental selenium, what can we do to improve the selenium status of dairy cattle? One potential approach would be to use forages and grains with a higher selenium content. Plants, marine algae, and bacteria can convert inorganic selenium from the soil into organic selenoamino acids, like selenomethionine. These organic selenium sources are more available to the animal for absorption and utilisation. However, this approach would require a consistent source of high selenium feedstuffs and monitoring of selenium content would be time-consuming and costly.

One way that dairymen have been circumventing the selenium regulations is by strategically using selenium injections. These injectable forms of selenium avoid the problem of poor absorption due to the ruminal environment and are routinely administered to dairy cows during the dry period. This practice has been used for many years as a short-term therapy, but does very little to improve the long-term selenium status of the animal (Pavlata *et al.*, 2001). A study that looked at various storage tissues, found that injectable selenium was no better at improving selenium stores than the no Se group. In fact, only the group that received the organic selenium source, Sel-Plex®, showed significant improvements (Figure 6).

Maas and coworkers (1993) found that after an animal was injected with inorganic selenium, Se peaked at 5 hours post-injection. By 28 days, blood selenium in cows was around 50 ng/ml compared to the 100 ng/ml that was experienced during the first 24 hour period. So it appears that while injections may work as a short-term therapy, their effectiveness on body stores of selenium is very limited. Finally, injectable forms of selenium have been known to contribute to spontaneous abortions in the dry cow pen and are responsible for many injection site abscesses in the animals that receive them.

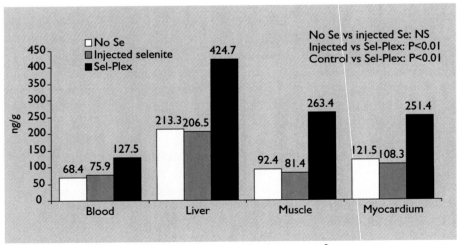

Figure 6. Blood and tissue Se of calves: injected Se vs. Sel-Plex®.

The major advantage of an organic selenium source over inorganic selenium source is its improved absorption and retention in the body. Selenoamino acids incorporated into body proteins provide a reserve of stored selenium when demand for selenium is high, particularly during disease challenge and gestation (Gunter *et al.*, 2003). Maternal transfer of selenium (through the placenta, colostrum and milk) improves the ability of the calf to survive and thrive (Table 4).

Table 4. Sel-Plex® improves Se status of cows and suckling calves.

	No Se	NaSe	Sel-Plex®	Probability of no effect	
				No Se vs. Se	NaSe vs. SP
Whole blood Se, ng/ml (cows)					
At calving	108	142	174	0.003	0.01
At 2 mo postpartum	93	150	187	0.003	0.03
Whole blood Se, ng/ml (calves)					
At birth	105	134	203	0.01	0.02
At 4 months	51	66	122	0.06	0.05

References

Arnér, E.S.J. and A. Holmgren, 2000. Physiological functions of thioredoxin reductase. *European Journal of Biochemistry* **267**:6102-6109.

Arechiga, C.F., O. Ortiz and P.J. Hanson, 1994. Effect of prepartum injection of vitamin E and selenium on postpartum reproductive function of dairy cattle. *Theriogenology* **41**:1251-1258.

Arthur, J.R., F. Nicol and G.J. Beckett, 1993. Selenium deficiency, thyroid hormone metabolism, and thyroid hormone deiodinases. *American Journal of Clinical Nutrition* **57**:236S-239S.

Conrad, H.R. and A.L. Moxon, 1979. Transfer of dietary selenium to milk. *J. Dairy Sci.* **62**:404-411.

Cristaldi, L.A., L.R. McDowell, C.D. Buergelt, P.A. Davis, N.S. Wilkinson and F.G. Martin, 2005. Tolerance of inorganic selenium in wether sheep. *Small Ruminant Research* **56**(1-3):205-213.

Davis, P.A., L.R. McDowell, N.S. Wilkinson, C.D. Buergelt, R. Van Alstyne, R.N. Weldon and T.T. Marshall. 2006a. Tolerance of inorganic selenium in range-type ewes during gestation and lactation. *J. Anim. Sci.* **84**:660-668.

Davis, P.A., L.R. McDowell, N.S. Wilkinson, C.D. Buergelt, R. Van Alstyne, R.N. Weldon and T.T. Marshall, 2006a. Tolerance of inorganic selenium in range-type ewes during gestation and lactation. *J. Anim. Sci.* **84**:660-668.

Davis, P.A., L.R. McDowell, N.S. Wilkinson, C.D. Buergelt, R. Van Alstyne, R.N. Weldon and T.T. Marshall, 2006b. Effects of selenium levels in ewe diets on selenium in milk and the plasma and tissue selenium concentrations of lambs. *Small Ruminant Research* **65**: 14-23.

Grace, N.D., 1994. Selenium. In: *Managing Trace Element Deficiencies*. New Zealand Pastoral Agricultural Research Institute Limited, Palmerston North, New Zealand, pp. 9-23.

Grace, N.D., J. Lee, R.A. Mills and A.F. Death, 1997. Influence of selenium status on milk selenium concentrations in dairy cows. *New Zealand Journal of Agricultural Research* **40**:229-232.

Gerloff, B.J., 1992. Effect of selenium supplementation on dairy cattle. *J. Anim. Sci.* **70**:3934-3940.

Gunter, S.A., P.A. Beck and J.M. Phillips, 2003. *Effects of supplementary selenium source on the performance and blood parameters in beef cows and their calves.* Sel-Plex Technical Report 189, Alltech, Inc., Nicholasville, KY.

Gupta, U.C. and J.H. Watkinson, 1985. Agricultural significance of selenium. *Outlook on Agriculture* **14**:183-189.

Harrison, J.H. and H.R. Conrad, 1984. Effect of selenium intake on selenium utilization by the nonlactating dairy cow. *J. Dairy Sci.* **67**:219.

Hemler, M.E., H.W. Cook and W.E.M. Lands, 1979. Prostaglandin biosynthesis can be triggered by lipid peroxidases. *Archives of Biochemistry and Biophysics* **193**: 340-345.

Hemler, M.E. and W.E.M. Lands, 1980. Evidence for a peroxide-initiated free radical mechanism of prostaglandin biosynthesis. *Journal of Biological Chemistry* **225**:6253-6261.

House, W.A. and A.W. Bell, 1994. Sulfur and selenium accretion in the gravid uterus during late gestation in Holstein cows. *Journal of Dairy Science* **77**:1860-1869.

Jacques, K.A., 2002. Selenium metabolism in animals – the relationship between dietary selenium form and physiological response. *Feed Compounder*.

Jukola, E., J. Hakkarainen, J. Soloniemi and S. Sankari, 1996. Blood selenium, vitamin E, vitamin A and B-carotene concentrations and udder health, fertility treatments and fertility. *J. Dairy Sci.* **79**:838-845.

Maas, J., J.R. Peanroi, T. Tonjes, J. Karlonas, F.D. Galey and B. Han, 1993. Intramuscular selenium administration in selenium deficient cattle. *J. Vet. Intern. Med.* **7**:342-348.

Macky, S.M., 1991. Selenium in dairy cattle - A practitioners approach. *Proceedings of the Society of Dairy Cattle Veterinarians of the New Zealand Veterinary Association* **8**:5-15

Marshall, P.J., R.J. Kulmacz and W.E.M. Lands, 1987. Constraints on prostaglandin biosynthesis in tissues. *Journal of Biological Chemistry* **262**: 3510-3517.

Maus, R.W., F.A. Martz, R.L. Belyea and W.F. Weiss, 1980. Relationship of dietary selenium to selenium in plasma. *J. Dairy Sci.* **63**:532-537.

National Research Council, 2001. *Nutrient Requirements of Dairy Cattle.* Seventh Revised Edition. National Academy Press, Washington, DC.

Oldfield, J.E., 2002. *Selenium World Atlas.* Selenium-Tellurium Development Association, Grimbergen, Belgium (pub.).

Ortman, K. and B. Pehrsen, 1997. Selenite and selenium yeast as feed supplements for dairy cows. *J. Vet Medicine A* **44**:373-380.

Pavlata, L., J. Illek and A. Pechova, 2001. Blood and tissue selenium concentrations in calves treated with inorganic or organic selenium compounds – A comparison. *Acta Vet. Brno* **70**:19–26.

Pehrson, B., 1993. Selenium in nutrition with special reference to the biopotency of organic and inorganic selenium compounds in Biotechnology in the Feed Industry. In: *Proceedings of the 9th Alltech Symposium.* Lyons, T.P. (ed.), Nottingham University Press, Nottingham, pp. 71-89.

Reddanna, P., J. Whelan, J.R. Burgess, M.L. Eskew, G. Hildenbrandt, A. Zarowker, R.W. Scholz and C.C. Reddy, 1989. The role of vitamin E and selenium on arachidonic acid oxidation by way of the 5-lipoxygenase pathway. In: Diplock, A.T., Machlin, L.J., Packer, L. and Pryor, W.A. (eds.). *Vitamin E: Biochemistry and Health Implications.* New York: New York Academy of Science, pp. 136-145.

Schwarz, K. and C.M. Foltz. 1957. Selenium as an integral part of factor 3 against dietary necrotic liver degeneration. *J. Am. Chem. Soc.* **79**:3242-3243.

Segerson EC, Riviere GJ, Bullock TR, Thimaya S, Ganapathy SN. 1980. Uterine contractions and electrical activity in ewes treated with selenium and vitamin E. *Biology of Reproduction* **23**:1020-1028.

Silvestre, F.T., D.T. Silvestre, J.E.P. Santos, C. Risco, C.R. Staples and W.W. Thatcher, 2006. Effects of selenium (Se) sources on dairy cows. *J. Anim. Sci.* **89**: (Suppl 1.) 52.

Smith, K.L., J.S. Hogan and W.P. Weiss. 1997. Dietary vitamin E and selenium affect mastitis and milk quality. *J. Anim. Sci.* **75**:1659-1665.

Spallholz JE, Boylan ML, Larsen HS. 1990. Advances in understanding selenium's role in the immune system. *Annals of the New York Academy of Science* **587**: 123-139.

Ullrey DE. 1992. Basis for regulation of selenium supplements in animal diets. *Journal of Animal Science* **70**: 3922-3927.

Van Saun, R. J. 1990: Rational approach to selenium supplementation essential. *Feedstuffs* (Jan 15): 15-17.

Weiss, W.P., J. S. Hogan, K.S. Smith, and K.H. Hoblet. 1990. Relationships among selenium, vitamin E, and mammary gland health in commercial dairy herds. *J. Dairy Sci.* **73**:381-390.

Weiss, W. P. 2003. Selenium nutrition of dairy cows: comparing responses to organic and inorganic selenium forms. In: *Biotechnology in the Feed Industry: Proceeding from Alltech's 19 th Annual Symposium* (T.P. Lyons and K.A.Jacques, eds), Nottingham Press University, Nottingham, UK, pp. 333-343.

Wichtel J.J., A.L. Craigie, D.A. Freeman, H. Varela-Alvarez and N.B. Williamson, 1996. Effect of selenium and iodine supplementation on growth rate and on thyroid and somatotropic function in dairy calves at pasture. *J. Dairy Sci.* **79**:1866-1872.

Wright PL, Bell MC. 1966.Comparative metabolism of selenium and tellurium in sheep and swine. *American Journal of Physiology* **211**: 6-10.

Raising replacement calves: nutrition, management and objectives

Alex Bach
Nutrition, Management, and Welfare Research Group, IRTA, Spain

1. Introduction

Current intensive dairy production systems suffer from reproductive problems and relatively short productive lives of dairy cows. Therefore, producing quality replacement heifers, with the least mortality and greatest efficiency, is gradually becoming more important. However, replacement heifers are expensive to raise, labour intensive, and there is some uncertainty about the optimum target weights, nutrition, and management.

The most common objective of raising Holstein dairy heifers is to have a first-calf heifer at about 22 months weighing about 635 kg (or about 570 kg post-calving) before calving and be 137-139 cm tall at the withers. Considering that Holstein calves are born with about 40 kg of BW, then an overall ADG of about 800 g/d should be targeted. However, this overall average can be achieved with many different growth curves (i.e. growing slowly early in life, then fast, then slow again...).

2. Enhanced growth feeding programs

The most common practice for raising replacement dairy calves consists on feeding 4 l/d of MR at a dilution rate of 12.5% DM from 3 to 60 d of age to attain ADG around 500 g/d. The aim of this conventional feeding program is to promote an early consumption of calf starter and to achieve a good rumen development while decreasing nutrition costs. However, achieving greater growth rates at early stages in life (2-3 months) might be profitable because increases in relative body weight (BW) and wither height (WH) are most rapid and cost-efficient during the first 6 months of life (Kertz *et al.*, 1998). Recent studies have shown that ADG of 0.78 kg/d can be obtained when feeding milk *ad libitum* (Jasper and Weary, 2002), or growth rates of 1 kg/d when feeding milk or MR at an increased rate of BW (DM from MR at 0.75 to 4% of BW) (Lodge and Lister, 1973; Diaz *et al.*, 2001).

In enhanced-growth feeding programs, high levels of MR and crude protein (CP) are recommended. In calves fed isoenergetic MR with increasing CP levels (from 14% to 26%) (Blome *et al.*, 2003), or in calves fed a MR with 30% CP at a dilution rate of 15% or 18% (Diaz *et al.*, 2001), growth rates and N retention (g/d) increased linearly as the MR protein:energy ratio and the MR dilution rate increased, respectively. However, plasma urea concentration also raised as the protein:energy ratio increased, indicating that calves fed MR at high CP did not utilise dietary N as efficiently as calves fed MR at low CP content (Blome *et al.*, 2003). In the study by Diaz *et al.* (2001), plasma urea N and efficiency of protein utilisation suggested that protein was not limiting growth. Thus, the efficiency of utilisation of absorbed amino acids (AA) could be improved by lowering the CP content of MR, and adjusting its amino AA profile. In fact, Kanjanapruthipong (1998) reported that calves fed MR (concentrated at 13% DM and 21% CP) containing soy protein and supplemented with AA presented a greater ADG and N retention (g/d) than calves receiving a MR without AA supplementation. A recent study (Terré *et al.*, 2006d) has suggested that conventional feeding programs (4 l of MR/d) might be limiting growth due to an insufficient supply of Met, Lys, His, and Trp.

On the other hand, young calves achieve mature ruminal function within a few weeks after dry feed is first offered (Lalles and Poncet, 1990). Therefore, early dry feed consumption results in a greater rumen metabolic activity in calves (Anderson *et al.*, 1987). Because calves following an enhanced-growth feeding program usually have low starter intakes during the preweaning period (Shamay *et al.*, 2005), it might be expected that this type of feeding programs act negatively on the establishment of the rumen function. Young calves achieve mature ruminal function 2-3 weeks after dry feed is first offered (Lallès and Poncet, 1990). Therefore, early dry feed consumption improves early rumen microbial development, resulting in a greater rumen metabolic activity (Anderson *et al.*, 1987). Then, the high level of milk replacer in calves following an enhanced-growth feeding program, may delay the start of dry feed consumption, and consequently it may delay rumen development (Davis and Drackley, 1998). In fact, duodenal microbial flow of calves following an enhanced-growth feeding program was lower than that of calves fed conventionally (Terré *et al.*, 2006c), which may explain the observations that calves on an enhanced-growth feeding program had lower dry matter intake during the preweaning and postweaning periods, and lower apparent nutrient digestibility coefficients than calves reared conventionally the week after weaning (Terré *et al.*, 2006c).

3. Management

A common pitfall reported in all studies using enhanced-growth feeding programs is that calves struggle at weaning, and average ADG during the week after weaning is reduced more than half compared with the ADG achieved before weaning (Bar-Peled *et al.*, 1997; Jasper and Weary, 2002; Terré *et al.*, 2006a), probably as a consequence of the low starter intake during the preweaning period. Lambs seem to learn the eating behaviour through the sight of the adult behaviour (Phillips and Youssef, 2003). Similarly, calves reared in groups increased fresh grass intake compared with calves reared individually (Phillips, 2004). The author suggested that the sight of feed being taken into the mouth was the relevant stimulus to learn the feeding behaviour from each other. A recent study (Terré *et al.*, 2006a) shows that rearing calves in groups following an enhanced-growth feeding program does not play a role in increasing starter intake. Calves reared in groups performed cross-sucking and inter-sucking behaviours, but there were no differences in days on medical treatments or with scours, and both calves performed similarly throughout the study.

4. And now, do it without antibiotics...

The use of antibiotics in milk replacer (MR) of calves has been a common practice in animal production to improve feed efficiency and prevent diseases, especially scours during the first weeks of calves life. However, the overuse of antibiotics exerts selective pressure that renders antibiotics ineffective (Amabile-Cuevas *et al.*, 1995), and EU banned the use of antibiotics as growth promoters in animal nutrition (1831/2003 EEC). Several additives have been proposed to improve calf health as alternatives to the use of antibiotics as growth promoters. Mannan oligosaccharides (MOS) are complex mannose sugars derived from cell wall fragments of yeast that are believed to block the colonisation of digestive pathogens increasing the competition for attachment sites in the digestive tract (Heinrichs *et al.*, 2003). When MR was supplemented with antibiotics or MOS, a reduction in calf scours was observed compared with control (no additives) treatment (Heinrichs *et al.*, 2003). Furthermore, feeding fructo ologosaccharides in combination to spray-dried bovine serum to calves reduced the incidence and severity of enteric disease (Quigley *et al.*, 2002). However, MOS supplementation in the MR or concentrates have shown inconsistent results. Calves fed MR with 4 g/d of MOS increased starter intake, but this increase did not result in growth differences (Heinrichs *et al.*, 2003). Contrarily, calves receiving a combination of probiotics, allicin, and fructoologosaccharides resulted in equivalent performance than calves that received antibiotics during the first 5 weeks of life (Donovan *et al.*, 2002).

Although MOS have been reported to alter lymphocyte response in vitro (Muchmore *et al.*, 1990), its effects on animal immune system are not well established. In recent years, enhanced-growth feeding programs for dairy calves have been proposed to increase growth rate during the pre-weaning period (Brown *et al.*, 2003; Shamay *et al.*, 2005). Feeding high amounts of milk replacer (Diaz *et al.*, 2001; Jasper and Weary, 2002) resulted in increased faecal consistency. Thus, the use of MOS could help to palliate any negative effects of increasing faecal consistency, avoiding the colonisation of digestive pathogens, and improve feed efficiency of milk replacer.

A recent study (Terré *et al.*, 2006b) showed that MOS supplementation increases starter intake and may reduce the incidence of *Cryptosporidium* spp.

References

Amabile–Cuevas, C., M. Cardenas-García and M. Ludgar, 1995. Antibiotic resistance. *American Scientist* **83**:320-329.

Anderson, K.L., T.G. Nagaraja, J.L. Morrill, T.B. Avery, S.J. Galitzer and J.E. Boyer, 1987. Ruminal microbial development in conventionally or early-weaned calves. *J. Anim. Sci.* **64**:1215-1226.

Bar-Peled, U., B. Robinzon, E. Maltz, H. Tagari, Y. Folman, I. Bruckental, H. Voet, H. Gacitua and A.R. Lehrer, 1997. Increased weight gain and effects on production parameters of Holstein heifer calves that were allowed to suckle from birth to six weeks of age. *J. Dairy Sci.* **80**:2523-2528.

Blome, R.M., J.K. Drackley, F.K. McKeith, M.F. Hutjens and G.C. McCoy, 2003. Growth, nutrient utilization, and body composition of dairy calves fed milk replacers containing different amounts of protein. *J. Anim. Sci.* **81**:1641-1655.

Davis, C.L. and J.K. Drackley, 1998. *The development, nutrition, and management of the young calf*, first edition. Iowa State University Press, Iowa.

Diaz, M.C., M.E. Van Amburgh, J.M. Smith, J.M. Kelsey and E.L. Hutten, 2001. Composition of growth of Holstein calves fed milk replacer from birth to 105-kilogram body weight. *J. Dairy Sci.* **84**:830-842.

Donovan, D.C., S.T. Franklin, C.C.L. Chase and A.R. Hippen, 2002. Growth and health of Holstein calves fed milk replacers supplemented with antibiotics or enteroguard. *J. Dairy Sci.* **85**:947-950.

Heinrichs, A.J., C.M. Jones and B.S. Heinrichs, 2003. Effects of mannan oligosaccharide or antibiotics in neonatal diets on health and growth of dairy calves. *J. Dairy Sci.* **86**:4064-4069.

Jasper, J. and D.M. Weary, 2002. Effects of ad libitum milk intake on dairy calves. *J. Dairy Sci.* **85**:3054-3058.

Kanjanapruthipong, J., 1998. Supplementation of milk replacers containing soy protein with threonine, methionine, and lysine in the diets of calves. *J. Dairy Sci.* **81**:2912-2915.

Kertz, A.F., B.A. Barton and F. Reutzel, 1998. Relative efficiencies of wither height and body weight increase from birth until first calving in Holstein cattle. *J. Dairy Sci.* **81**:1479-1482.

Lallès, J.P. and C. Poncet, 1990. Changes in ruminal and intestinal digestion during and after weaning in dairy calves fed concentrate diet containing pea or soya bean meal. 1. Digestion of organic matter and nitrogen. *Livest. Prod. Sci.* **24**:129-142.

Quigley, J.D. III, C.J. Kost and T.A. Wolfe, 2002. Effects of spray-dried animal plasma in milk replacers or additives containing serum and oligosaccharides on growth and health of calves. *J. Dairy Sci.* **85**:413-421.

Shamay, A., D. Werner, U. Moallem, H. Barash and I. Bruckental, 2005. Effect of nursing management and skeletal size at weaning on puberty, skeletal growth rate, and milk production during first lactation of dairy heifers. *J. Dairy Sci.* **88**:1460-1469.

Terré, M., A. Bach, and M. Devant, 2006a. Performance and behavior of calves reared in groups or individually following an enhanced-growth feeding program. *J. Dairy Res.* **73**:480-486.

Terré, M., M.A. Calvo, C. Adelantado, A. Kocher and A. Bach, 2006b. Effects of mannanoligosaccharides on performance and microorganism fecal counts of calves following an enhanced-growth feeding program. *Anim. Feed Sci. Technol.* In press.

Terré, M., M. Devant and A. Bach, 2006c. Effect of level of milk replacer fed to Holstein calves on performance during the preweaning period and starter digestibility at weaning. *Livestock Sci.* In press (corrected proof).

Terré, M., M. Devant and A. Bach, 2006d. Performance and nitrogen metabolism of calves fed conventionally or following an enhanced growth feeding program during the preweaning period. 2006. *Livestock Sci.* **103**:109-119.

Managing optimal health of dairy calves

Arlyn Judson Heinrichs
The Pennsylvania State University, University Park, Pennsylvania, USA

1. Introduction

Raising dairy calves from birth to weaning in a healthy, economical, and responsible manner depends on a large number of factors. These include the combination of a healthy dam that calves without difficulty, a clean calving area, the early ingestion of adequate quantities of good-quality colostrum with adequate passive transfer of colostral immunoglobulins to the calf, comfortable and adequate housing, and adequate nutrition following the colostral feeding period up to the time of weaning. Calf morbidity and mortality are perennial problems in all countries where cattle are raised. Most illnesses and deaths occur in the first few weeks of life because of the effects of infection pressure, lack of sufficient colostral immunity, and inadequate housing, or the effects of adverse environmental conditions.

Care of the dry cow prior to calving, neonatal behaviour, immunoglobulin absorption, nutritional and feeding requirements, and various housing strategies have all been studied under experimental conditions (Waltner-Toews *et al.*, 1986f). In spite of the many studies that have been done, the effectiveness of calf raising programs in reducing calf morbidity and mortality has not been achieved. There is however, sufficient knowledge available that can be applied at the farm level to manage the health and production of dairy calves so that morbidity and mortality can be maintained at low levels. Rearing dairy herd replacement animals represents a long-term investment in feed, labour, and other resources to ensure high-quality replacements for the lactating herd (Clark *et al.*, 1984). Through cooperative research, problems involving nutrition, housing, and management for replacement animals have been studied (Losinger and Heinrichs, 1997).

2. Dairy calf mortality and morbidity incidence

The infectious diseases of the digestive and respiratory tracts are the most critical diseases of calves from birth up to several months of age. Approximately 75% of the mortality of dairy animals less than one year of age occurs during the first month of life (Losinger and Heinrichs, 1996). This indicates the necessity of giving high priority to a health management system for rearing newborn calves, especially during the first month of life. Recent United States Department of Agriculture data (USDA, 2002) showed that in the United States calf mortality is 8.7% with scours and respiratory representing the major causes of mortality.

Abortions, stillbirths, and congenital defects account for approximately two to 3% of calf mortality and include the common causes of abortion, such as brucellosis, leptospirosis, and others, intrapartum hypoxemia caused by prolonged parturition, and inherited and non-inherited congenital defects. Acute diarrhoea accounts for approximately 62% of the mortality of dairy calves less than three weeks of age. The most important pathogens associated with diarrhoea are enterotoxigenic *Escherichia coli* in calves under three to five days of age, rotavirus in calves seven to ten days of age, corona-virus in calves seven to 15 days of age, *Cryptosporidia* spp. in calves 15 to 35 days of age, *Salmonella* spp., usually in calves several weeks of age, and coccidiosis (*Eimeria* spp.) in calves older than three weeks of age. Coliform septicaemia caused by invasive strains of *E. coli* occurs most commonly in calves under four days of age that are agammaglobulinemic because of the failure of passive transfer of colostral immunoglobulins. Enzootic Pneumonia occurs primarily in housed calves over two months of age. It is caused by infection with respiratory viruses and inadequate ventilation, and accounts for about 21% of calf mortality from birth to six months of age (USDA, 2002).

The development and application of the principles of dairy calf health and production management require some knowledge of the descriptive epidemiology of calf morbidity and mortality, achievable targets of performance, risk factors associated with calf morbidity and mortality, and the possible long-term effects of calfhood disease on subsequent performance. In addition, a system of monitoring calf health and production must be established and used to analyse performance and to make recommendations for improvement.

Dairy calf morbidity statistics are not as reliable as those on mortality because they depend on the producer's clinical diagnosis, whether the animal was treated for the illness, and the tendency of producers not to record every illness event. On the other hand, mortality data are more readily available because it is easier to count dead calves. In a survey of dairy calf morbidity and mortality in Holstein dairy herds, 20% of live-born calves were treated for diarrhoea and 15% were treated for pneumonia before weaning (Waltner-Toews *et al.*, 1986f). In a study of 1171 Holstein heifer calves born over a two-year period in 26 commercial dairy herds in New York State, the crude incidence rates of diarrhoea within

14 days of birth, diarrhoea from 15 to 90 days of age, dullness, and respiratory illness and death were 9.9, 5.2, 7.7, 7.4 and 3.5 per 100, respectively, for the period of study (Curtis *et al.*, 1988a). Data from documented field observations in a large farm study showed that while average mortality was 0.5%, the range in this 21 farm study was from 0 to 18.8% during a 12 month period (Place *et al.*, 1998), revealing the variability that exists within a population of farms.

A study of 26 dairy herds in New York by Curtis *et al.* (1988a,b) indicated that management directly affected the risk of respiratory illness within 14 days of birth. The environment in which the calf is raised also has a profound effect on health and growth. Pritchard *et al.* (1981) found that treatment for respiratory disease and lung infections in veal calves was directly related to daily weight gains and was associated with air quality in the animal unit.

3. Long-term effects of calf morbidity on health and performance

The effect of early calfhood health status on survivorship and age at first calving (AFC), after controlling for the farm effect, has been examined (Waltner-Toews *et al.*, 1986a). Heifers that had been treated for pneumonia during the first three months of life were 2.5 times more likely to die after 90 days of age than heifers that had not been treated. Heifers with a calfhood history of being treated for diarrhoea were 2.5 times more likely to be sold for dairy purposes than other calves, and heifers that had been treated for diarrhoea were 2.9 times more likely to calve after 900 days (30 months) of age than other heifers (Waltner-Toews *et al.*, 1986a). Roy (1983) has suggested that, if abortions are excluded, calf mortality can be subdivided into perinatal mortality (stillbirths at >270 days of gestation and mortality during the first 24 hours of life), neonatal mortality (calves born alive that die between 24 hours and 28 days) and older calf mortality (calves born alive that die between 29 and 182 days). One of the few studies to measure individual dairy calf mortality was done in Canada (Waltner-Toews *et al.*, 1986f); mortality was 3.8% for all calves, with a herd median mortality of 0% (minimum to maximum, 0 to 67%). In a survey of 407 dairy herds in Virginia reporting 12,300 heifer calf births over a six-month period, the mortality at birth averaged 1.2% and the mortality of calves born alive and up to three months of age averaged 6.5% (James *et al.*, 1984). In a one-year survey of Norwegian dairy herds, the stillbirth rate and calves dying within 24 hours of birth was 2.80%, and the overall mortality rate to 30 days was 3.95%, which is extremely low compared to other countries (Simensen, 1982). In five Libyan dairy stations, each of which contained 500 dairy cows, over a period of five years, the calf mortality rate from birth to 30 days ranged from 12.5 to 26% for the various stations (Gusbi and

Hird, 1983). The overall calf mortality rate from birth to 90 days was 18.8%. In a study of Irish dairy herds over a nine-year period, with a total of 9,430 calvings, the average incidence of abortion was 1.8%, perinatal mortality (stillbirths and deaths within 48 hours) was 8.37%, and 73% were dead at birth (Mee, 1988). In well-managed herds, the annual mortality rate of calves under 30 days of age can be controlled at a level below three to 5% of all calves born alive and normal. While these data are quite dated, available USDA data shows that these numbers have not changed substantially in past years (USDA, 2003).

While the economic losses resulting from calfhood morbidity and mortality are recognised by the dairy industry, the possible long-term effects of morbidity on health and performance, although unknown, may constitute a loss of even greater economic importance (Curtis *et al.*, 1989). It has been suggested that calves that survive clinical episodes of disease might be affected by some longer residual effects on growth, age at calving, body weight at calving, reproductive function, and milk production (Waltner-Toews *et al.*, 1986e; Place *et al.*, 1998). Calfhood diseases may affect feed intake, particularly dry feeds, depending on the type and severity of the disease, and the effect may last for up to several days or a few weeks beyond the course of the illness. Such a reduction in feed intake, combined with the effects of the lesion, could be expected to have an effect on subsequent survivability and productivity (Waltner-Toews *et al.*, 1986e). Some of these effects have now been measured under field conditions, with variable results. Based on the analysis of the records of 34 Ontario Holstein dairy herds, the effects of diarrhoea and pneumonia in calves on their subsequent survivorship and age at first calving was assessed. Heifers that had been treated for pneumonia during the first three months of life were 2.5 times more likely to die after 90 days of age than heifers that had not been treated for pneumonia, after controlling for the farm effect. Heifers with a calfhood history of being treated for diarrhoea were 2.5 times more likely to be sold for dairy purposes than other calves (Waltner-Toews *et al.*, 1986e). Heifers that had been treated for diarrhoea were 2.9 times more likely to calve after 900 days of age than other heifers, after controlling for the farm effect.

The epidemiologic determinants or risk factors of disease and suboptimal performance that are important in dairy calves under one month of age include the vigour and health of the calf at birth, the level of colostral immunity achieved in the calf, management practices of the personnel caring for the calves, nutritional program, including the quality of milk replacers (if used), herd size, the adequacy of the housing and environment, including the hygiene of the calf's microenvironment, and the season of the year (Simensen, 1986;

Hancock, 1983). Path analysis of individual calf risk factors for dairy calfhood morbidity and mortality in New York dairy herds indicated that management appears to affect the risk of respiratory disease directly and indirectly (Curtis *et al.*, 1988b). Being born in loose housing increases the risk of diarrhoea within 14 days of birth and from 15 to 90 days of birth. Calves of first-calf heifers are at increased risk of respiratory illness and calves whose dams were vaccinated against *E. coli* are at decreased risk of death. Calves with diarrhoea within 14 days of birth or calves with dullness and diarrhoea from 15 to 90 days of age are at increased risk of respiratory illness. Calves that are dull and listless, with droopy ears, and off feed within 90 days of birth are at the greatest risk of death.

In a large study, Curtis *et al.* (1989) followed 1,171 Holstein heifer calves on several New York dairy farms. Their findings yielded incidence rates for scours of 9.9% within 14 days of birth, 5.2% from 15 to 90 days of age, 7.7% for calves displaying dullness, and 7.4% for calves with respiratory illness. This study was followed by Correa *et al.* (1988), who evaluated the effects of calf morbidity on age at first calving (AFC) on the same animals. Heifers without respiratory illness as calves were twice as likely to calve and calved six months earlier compared to those with respiratory illness as calves. An unexpected result from this study was that heifers displaying dullness or unthriftiness as calves were 1.6 times more likely to calve and calved two months earlier when compared to calves without dullness as calves. Dullness would be expected to increase AFC because of anticipated lower growth rates from inadequate feed intake and less active or normal behaviour.

Health status of dairy heifers has been shown to have a significant impact on growth rate of calves especially during the first six months of life (Donovan *et al.*, 1986). Season of birth and occurrence of diarrhoea, septicaemia, and respiratory disease can significantly decrease heifer growth (height and weight). Donovan *et al.* (1986) reported that these variables plus farm, birth weight, and exact age when six-month data are collected explained 20 and 31% of the variation in birth weight and pelvic height growth respectively from birth to six months. Septicaemia and pneumonia slowed growth by 13 to 15 days (to reach the same weight as healthy calves) during the first six months, while diarrhoea had a much smaller influence on growth (Donovan *et al.*, 1986). Passive transfer of colostral immunoglobulins had no direct effect on growth but did influence weight and height growth through its effect on health (Donovan *et al.*, 1986). Place *et al.* (1998) showed that housing and season had significant effects on average daily gain (ADG). Other variables, such as calving location, parity of the dam, and delivery score at calving, had significant effects on ADG to four months of age.

The effects of age and seasonal patterns, association of management with morbidity, and association of management with mortality in dairy herds in Ontario have also been studied (Waltner-Toews *et al.*, 1986b,c,d). The risk of diarrhoea peaked during the second week of life; the peak incidence of pneumonia occurred during the sixth week of life. Treatment rates for diarrhoea and pneumonia were generally lower in spring and summer than during the autumn and winter. The larger farms had a significantly greater chance of experiencing mortality than smaller farms in both winter and summer. Farms that had policies of attending calvings and ensuring that calves received their first colostrum had a significantly lower probability of experiencing winter mortality than farms that did not have these policies. Farms that housed calves in hutches had a lower chance, and those that housed calves in group pens had a higher likelihood of experiencing summer mortality than farms that used individual indoor calf pens.

Several factors can affect the vigour and health of the calf immediately after birth. If the nutritional status of the dairy female at the time of parturition is adequate, nutritional deficiencies at birth are usually rare. Calves born from dams affected with any disease associated with prolonged anorexia, fever, or septicaemia may be weak. Morbidity and mortality in dairy calves are associated with difficult calving in first-calf heifers. Dystocia is affected by several non-genetic and genetic factors, including age and parity of the dam, size of the calf, season of the year, and degree of birth difficulty. In some studies, up to 52% of first-parity dams received assistance in calving (Sieber *et al.*, 1989). Correlations between dystocia and calf mortality have been about 0.7 (Weller *et al.*, 1988). The genetic relationship between calf survival and calving difficulty of Holstein cattle has been examined (Martinez *et al.*, 1983a,b). The high correlation between calving ease and calf survival means that any resultant improvement in calving ease as a result of sire selection or selective breeding is expected to result in improvement of calf survival (Cue and Hayes, 1985). However, there is some evidence that cows with a maternal disposition to easy calving produce daughters with a disposition toward difficult calving. This negative, direct maternal relationship occurs because small calves are born with ease, the small calf becomes a small cow, and the small cow has difficulty calving at early parities (Thompson and Rege, 1984).

Other studies look at calf factors affecting future performance of the heifer. Place *et al.* (1998) found that factors that were related to increased age at first calving were increased difficulty of delivery, antibiotic treatment, nutrition before weaning, forage quality after weaning, and the environment in calf housing areas. Body weight at calving increased with parity of the dam, calf nutrition, and increased mean temperature of the calf housing area. Body condition score at calving appeared to be positively influenced by delivery score at calving, dam parity, and liquid feed intake. Withers height at calving was positively affected by treatment of animals with antibiotics and increased mean temperature in the calf area. This study shows that health events, the environment, and nutrition of the young calf before weaning have long-term effects on the growth of these animals up to calving.

4. Passive immunity transfer to the newborn calf

Calves are born agammaglobulinemic and the most important factors influencing the health of the newborn calf are the ingestion of liberal quantities of good-quality colostrum within a few hours after birth and the subsequent absorption of sufficient amounts of colostral immunoglobulins within the first 24 hours after birth, when closure occurs (Duhamel and Osburn, 1984). The colostral immunoglobulins that are absorbed provide protection against systemic infections, such as coliform septicaemia. The colostral immunoglobulins ingested after closure are not absorbed but remain in the lumen of the intestine and provide local or lactogenic immunity. This is particularly important for protection against the rotavirus and coronavirus infections that cause diarrhoea in calves after a few days of age.

Some of the absorbed and circulating immunoglobulin G_1 (IgG_1) is transferred back to the intestine and retains significant functional antibody activity in the intestine (Besser *et al.*, 1988a,b). This mechanism accounts for most of the IgG clearance. The concentration of colostral immunoglobulins declines rapidly by two to three days after parturition to low levels in the postcolostral milk, which makes calves susceptible to viral infections. The mortality rate resulting from diarrhoea is much higher in calves with low levels of serum colostral immunoglobulins than in calves with adequate levels. Partial or complete failure of passive transfer of colostral immunoglobulins is a major determinant of liability to neonatal disease and mortality in calves (Gay *et al.*, 1988; Odde, 1988). When considering the causes of failure of passive transfer of colostral immunoglobulins, the transfer process is affected by three major factors.

The calf must ingest an adequate total mass of immunoglobulin within a few hours after birth. The concentration of IgG_1 in the colostrum, age at first feeding, and the volume of colostrum are major factors affecting colostral IgG absorption (Stott and Fellah, 1983). Immunoglobulin G concentration in the serum of the calf at 24 hours after feeding colostrum has a positive linear relationship with the IgG concentration in the colostrum. However, some results suggest a physiologic limitation to the mass of immunoglobulin that can be absorbed to serum from a given volume of colostrum (Besser *et al.*, 1985). The calf must ingest an immunoglobulin mass of 80 to 100 g to achieve adequate serum levels of immunoglobulins.

The absorption of immunoglobulins by the intestine into the systemic circulation of the calf is another important factor. Colostral immunoglobulins are absorbed for up to 24 hours after birth in calves, but the maximum efficiency of absorption occurs during the first six to 12 hours after birth. This rapid decline in absorption combined with delayed ingestion of colostrum results in failure of passive transfer, which occurs in ten to 30% of calves, even under ideal conditions (Donovan *et al.*, 1986). The termination of absorption of colostral immunoglobulins in the calf occurs spontaneously with age at a progressively increased rate after 12 hours postpartum (Stott *et al.*, 1979).

Substantial transmission of IgG can occur as late as 24 hours after birth but, in heavily contaminated environments, it is vital that maternal antibody reaches the lumen of the digestive tract before infection occurs (Michanek *et al.*, 1989). The serum immunoglobulins of newborn calves may also vary according to the season of the year. Mean monthly serum IgG_1 concentrations may be lowest in calves born in the winter months (November to May) in temperate climates such as Washington State and Scotland, and increase in calves born in the spring and early summer to reach peak levels in September, after which they begin to decrease again (Gay *et al.*, 1983). The opposite has been described for calves born in warmer climates such as Florida, where the highest levels occurred in February and March and lowest levels during the warm summer months (Donovan *et al.*, 1986).

The effects of passive immunity in dairy calves on subsequent survival and productivity indicate that calves that receive an adequate amount of colostrum and absorb sufficient immunoglobulins perform better than calves with insufficient levels (Robison *et al.*, 1988; DeNise *et al.*, 1989). The serum concentrations of colostral immunoglobulins can be measured for prospective risk assessment of individual calves and for retrospective evaluation of sick calves. The regular monitoring of serum immunoglobulin levels in calves on a herd basis would provide an evaluation of the colostrum feeding program and would assist in determining the causes of excessive morbidity and mortality in calves (Hancock, 1985).

5. Calf exposure to pathogens after birth

In most cases, the calf is born without any pathogens in its body. However, within minutes after birth, it becomes contaminated with the wide variety of pathogens that are in the immediate environment (Ruckebusch *et al.*, 1983). It has always been assumed that the level of infectious disease pressure experienced by the newborn calf is directly proportional to the concentration of pathogens in the environment, which includes the bedding or ground surface, walls of the pens, air, and other environmental structures that the calf comes in contact with soon after birth. The dam of the calf and other animals in the immediate vicinity, particularly diarrheic animals or animals ill with other infections, constitute the main source of most pathogens. Calves born into a highly contaminated environment may become infected during birth with enteropathogens carried by their dams or other cows. Enteric pathogens are spread within a herd through the faeces of infected animals and all inanimate objects that can be contaminated by faeces, including bedding, pails, boots, tools, clothing, and feed and water supplies. Heavily used calving pens can become heavily contaminated with enteric pathogens, such as *E. coli*, *Salmonella* spp., rotavirus, coronavirus, *Cryptosporidium* spp. and *Eimeria* spp. Enterotoxigenic *E. coli* are relatively resistant to environmental factors and can survive for long periods under the right conditions. One serogroup of the organism survived as long as 6.5 months in calf crates contaminated with calf faeces (Acres, 1985). The total numbers of bacteria in a calf pen or barn continue to increase as newborn calves are introduced successively into the facility and the facility is continually occupied by animals without depopulation, a clean out, disinfection, and period of vacancy (Roy, 1980). This increased occupation time results in a build-up of infection that is associated with an increasing incidence of enteric disease and suboptimal performance, which can be resolved by moving the animals out of the facility, a clean out, and disinfection, followed by a period of vacancy for several days.

Enterotoxigenic *E. coli*, rotavirus, coronavirus, and *Cryptosporidium* spp., the four major causes of diarrhoea in calves, are responsible for 75 to 95% of enteric infections of calves worldwide (Tzipori, 1985). The relative frequency of each of these pathogens varies from farm to farm, year to year, season to season, and country to country. Infections with one pathogen or mixed infections may predominate (Pohjola *et al.*, 1986). The prevalence and incidence of enteropathogens in calves are also age-dependent. Enterotoxigenic *E. coli* (ETEC) affects calves primarily under two days of age (Acres, 1985). In contrast, rotavirus, coronavirus, and *Cryptosporidium* affect calves from three days up to four weeks (Tzipori, 1985). Although these pathogens may be necessary causes of the diarrhoea, they may not be sufficient causes. It is important to remember that these pathogens can also be found in healthy calves. In some surveys, the rotavirus can be found in the faeces of up to 50% of healthy calves (Reynolds *et al.*, 1986). In microbiologic surveys of diarrheic and healthy calves, it is common to find mixed infections in diarrheic calves, whereas in healthy calves it is unlikely to find three or more pathogens (Reynolds *et al.*, 1986). In a survey of the faecal samples of diarrheic calves on farms experiencing outbreaks of diarrhoea, the rotavirus was the most common enteropathogen found, being excreted by more than half the diarrheic calves (Snodgrass *et al.*, 1986).

6. Calf and heifer housing and management

The provision of a proper environment for calves, in addition to adequate nutrition and other health-related programs, is critical to the success of raising herd replacements. The diagnosis and treatment of sick calves has long been recognised, but the need for effective management of environment to ensure optimal calf health has received much less attention (Bates and Anderson, 1984). On most dairy farms, calves are taken from the maternity area soon after birth and placed in the calf-rearing part of the main barn. Calves raised for herd replacements are commonly maintained in facilities that range from total confinement in narrow, elevated, individual stalls, tethered along an inside wall of the main barn, group pens, and individual calf hutches to open pastures. The greater the degree of confinement, the more the calves' environment is under the control of the producer, and the less calves are able to make behavioural responses to that environment. Morbidity and mortality rates are usually higher in calves housed indoors than outdoors, but there may be no difference between the health of calves raised in individual pens in an insulated, heated barn and the health of calves raised in outdoor hutches. The increased illness and mortality in calves that are reared indoors is often attributed to a combination of inadequate control of the thermal environment, poor air quality, undesirable relative humidity, inadequate exchange of air, and poor sanitation. However, open sheds and hutches can also become contaminated with common calf pathogens and, when combined with environmental stressors, can result in disease.

Surveys of dairy calf and heifer housing have indicated that many housing practices are less than adequate (Heinrichs *et al.*, 1987a,b). Dairy producers also recognise calf housing as a problem that is important to the veterinarian, building designer, and manufacturer. Producers who recognise the problem are more receptive to logical and economical solutions through educational programs, new building designs, and equipment. Under controlled conditions, many different types of calf and heifer housing systems yield satisfactory results when proper management for that housing is undertaken.

The calf is a homeotherm and is capable of maintaining a relatively constant body temperature throughout a wide range of ambient temperatures. The internal core temperature of the calf is regulated by the animal's ability to achieve a balance between the heat it produces in metabolism and the heat gained or lost from the environment (Bickert and Herdt, 1985). The calf exchanges heat energy with its surroundings through conduction, convection, radiation, and evaporation. Changes in the immediate physical and thermal surroundings of the calf that affect heat exchange through one or more of its four modes can affect thermoregulation and increase susceptibility to disease. The essential environmental needs of the young calf are comfort, space, and hygiene (Webster, 1984). The thermal environment, which includes air temperature and air movement, must be comfortable, and the physical space such as the floors and surfaces with which the calf makes contact must be comfortable. The space available to each calf should at least give it sufficient room to stand up, lie down, turn around, stretch its limbs, and groom itself. It should also provide calves with a reasonable degree of social intercourse. The microbiologic environment consists of the floor and wall surfaces, the bedding used, the air quality, the contact the calf makes with other calves, and the utensils and materials used by personnel in feeding and handling the animals.

Many cases of poor health in dairy calves have been traced to inadequate management and ventilation in the calf nursery (Anderson, 1978). Calves reared indoors are commonly affected with pneumonia caused by viruses and bacteria, which may reach high concentrations in the air of poorly ventilated, damp, and cold calf barns. Calf barns that are overcrowded, dark, and damp in the winter months and hot and poorly ventilated during the summer months commonly predispose calves to diseases of the respiratory tract. Unfortunately, in most dairy herds, the emphasis is on the milking cow and the calf barn traditionally has received low priority, resulting in uncomfortable rearing conditions that predispose to infectious diseases. Some producers have had so much difficulty raising calves indoors that they have resorted to rearing young calves in calf hutches outdoors. Whereas considerable research and effort have been expended in elucidating the infectious causes of respiratory diseases in calves raised indoors, only a few studies have been done to develop economical and practical methods of providing an optimum environment for calves raised indoors.

In general, dairy calf mortality is minimised best by a comprehensive management program that is followed stringently (Kertz, 1977). Published studies have emphasised the interrelationships of a multitude of management factors to calf mortality. Most common diseases of dairy calves under one month of age cannot be totally prevented and control at an economical level must be the major objective. With good management and disease control techniques, mortality can be maintained economically at a level below three to 5% of live-born calves under one month of age. A prerequisite for effective, economical control of calf disease is to establish a simple recording system that the producer can understand and use. Effective monitoring systems have been developed to achieve this task (Heinrichs *et al.*, 2003).

From birth to about six months of age, the calf is subjected to different infection pressures. In the first few days of life, the calf is susceptible to enterotoxigenic *E. coli*. Between five and 15 days of age, and sometimes later, rotavirus and coronavirus infection are the most common causes of enteritis. Salmonellosis may occur at any age, but commonly occurs in calves from one to two months of age, and respiratory infections occur most commonly between two and six months of age. Therefore, every economically feasible management effort should be made to reduce the infection pressure on the calf from birth to several months of age.

The reduction of the degree of exposure to infectious agents begins with the birth of the calf, which should take place in a clean environment. The perineum and udder of dirty cows should be washed shortly before calving. The umbilicus of the calf should be swabbed with 2% tincture of iodine immediately after birth to control the entry of environmental pathogens. Well-bedded calving barn stalls, calving corrals in large herds, and pastures provide the ideal environment for the newborn calf. Regular sanitation and hygiene are of paramount importance in the calf nursery, especially during the first few weeks of life, when the calf is most susceptible to infectious disease.

Keeping dairy calves alive and healthy starts with the maternity area. Herds that use well-bedded box stalls for calving and for the calf during the colostral feeding period have consistently lower calf mortality. Maternity pens should be well-bedded and dry. A Michigan study showed a significant relationship between the dry matter content of the bedding and calf mortality. Calf mortality was significantly lower when the bedding in the maternity pen was dry when compared with a maternity pen with wet bedding. Straw is considered the ideal bedding; sawdust is questionable because it does not absorb equivalent amounts of moisture. Also, calves tend to eat sawdust, which may precipitate digestive upsets and serve as a vehicle for the entrance of pathogenic organisms into the digestive tract. The nonspecific resistance of the newborn calf is markedly influenced by the type of housing, the temperature of the calving facilities, the temperature of the calf barn, the person caring for the calves, and whether attendance and assistance are provided at birth (Heinrichs *et al.*, 2003).

7. Colostrum feeding management

The ingestion of liberal quantities of colostrum by the newborn calf within the first six hours after birth is the first and most important nutritional requirement of the newborn calf. Surveys of calf and heifer management practices in dairy herds have indicated that husbandry practices for the newborn dairy calf are deficient in many aspects (Heinrichs *et al.*, 1987a,b; USDA, 2002). The timing and amount of colostrum feeding can be problematic. Field studies have indicated that if dairy calves are left with their dams, up to 25% may not suck within eight hours, and ten to 25% do not get an adequate amount of colostrum. The prevalence of failure of passive transfer in dairy herds could be minimised by artificially feeding all calves large volumes (three to four litres) of fresh or refrigerated first-milking colostrum within the first eight hours of life. Absorption of colostral immunoglobulins ceases by 24 hours after birth, but the continued ingestion of colostrum containing adequate levels of specific antibody provides local intestinal immunity against acute undifferentiated diarrhoea for as long as the colostrum is fed. Colostrum that is preserved by freezing is practical, economical, and reliable and may assist in reducing the incidence of infectious diarrhoeas because of the presence of colostral immunoglobulins, which provide local immunity (Foley and Otterby 1978). The continued presence of colostrum in the intestinal tract also exerts a local protective action against the infectious enteritides. The details of the availability, storage, treatment, composition, and feeding value of surplus colostrum have been published (Foley and Otterby, 1978). Recent work with mannan oligosaccharides have shown similar responses when fed to pre-weaned calves (Heinrichs *et al.*, 2003).

8. Monitoring calf and heifer performances

The production potential of heifers is a result of all previous breeding and management decisions. Many dairy farms follow approved calf and heifer health management practices, but several aspects of management need improvement (Heinrichs *et al.*, 1987a). The most important of these include the care and feeding of the newborn calf and the types and amounts of feeds fed to young calves, both of which have been shown to be related to high calf mortality.

Surveys of management practices on dairy farms have revealed a lack of systematic understanding of factors that are vital in calf health management (Heinrichs *et al.*, 1987a; Goodger and Theodore, 1986). In spite of the long-known importance of colostrum, a significant percentage of producers do not appreciate the important factors that influence the passive transfer of colostral immunoglobulins in calves and what is required to ensure a high level of colostral immunity in the calves. Colostrum intake is often left to the calf, and most producers are unsure of the amount of colostrum obtained by the calf. Some producers vaccinate pregnant cows to prevent infectious diarrhoea of calves but do not make the effort necessary to ensure that the calves ingest or are force-fed a sufficient amount of colostrum early enough after birth.

In many countries where grass is plentiful for several or more months of the year, young calves can be reared successfully on grass beginning shortly before or after being weaned from milk or milk replacer. Several factors are important for the successful rearing of young calves on pasture. The quantity and quality of grass available are of utmost importance because of the highly selective nature of grazing calves. Young calves can starve on unsuitable pastures unless they receive a supplemental concentrate. Digestibility trials have revealed that calves from three to five weeks of age can use approximately 75% of herbage dry matter. Compared with indoor feeding of hay and concentrates, pasture rearing has resulted in calves ruminating at an earlier age and developing greater rumen size and capacity. Pasture quality can also alter the response of calves to concentrate feeding. Calves on excellent pasture alone can make greater weight gains than calves on poorer pasture with supplementary grain feeding. Under conditions of poor pasture, low levels of nutrition, and adverse weather conditions, young calves are highly susceptible to parasitic gastroenteritis, which causes severe unthriftiness.

Systems need to be in place to monitor and evaluate the progress of heifers on dairy farms due to the large economic investment involved in raising dairy heifers (Tozer and Heinrichs, 2001). Since dairy replacements represent the future of the dairy herd it is imperative that they are well cared for in a manner that will allow them to optimise their genetic potential for the future of the farming operation. Optimal systems of rearing young dairy livestock are those that minimise time and financial investments and develop healthy, productive and timely replacement heifers for the milking herd.

References

Acres, S.D., 1985. Enterotoxigenic *Escherichia coli* infections in newborn calves: A review. *J. Dairy Sci.* **68**:229-256.

Anderson, J.F., 1978. Medical factors relating to calf health as influenced by the environment. *Bovine Pract.* **13**:3-5.

Bates, D.W. and J.F. Anderson, 1984. Environmental design for a total animal health care system. *Bovine Pract.* **19**:4-20.

Besser, T.E., A.E. Garmedia, T.C. McGuire and C.C. Gay, 1985. Effect of colostral immunoglobulin G_1 and immunoglobulin M concentrations on immunoglobulin absorption in calves. *J. Dairy Sci.* **68**:2033-2037.

Besser, T.E., C.C. Gay, T.C. McGuire and J.F. Evermann, 1988a. Passive immunity to bovine rotavirus infection associated with transfer of serum antibody into the intestinal lumen. *J. Virol.* **62**:2238-2242.

Besser, T.E., T.C. McGuire, C.C. Gay and L.C. Pritchett, 1988b. Transfer of functional immunoglobulin G (IgG) antibody into the gastrointestinal tract accounts for IgG clearance in calves. *J. Virol.* **62**:2234-2237.

Bickert, W.G. and T.H. Herdt, 1985. Environmental aspects of dairy calf housing. *Comp. Cont. Educ. Pract. Vet.* **7**:S309-S316.

Clark, A.K., D.L. Albright, L.D. Muller and F.G. Owen, 1984. Raising dairy replacement heifers. A review of NC-119 cooperative research. *J. Dairy Sci.* **67**:3093-3098.

Correa, M.T., C.R. Curtis, H.N. Erb and M.E. White, 1988. Effect of calfhood morbidity on age at first calving in New York Holstein herds. *Prev. Vet. Med.* **6**:253-262.

Cue, R.I. and J.F. Hayes, 1985. Correlation between calving ease and calf survival. *J. Dairy Sci.* **68**:958-962.

Curtis, C.R., H.N. Erb and M.E. White, 1988a. Descriptive epidemiology of calfhood morbidity and mortality in New York Holstein herds. *Prev. Vet. Med.* **5**:293-307.

Curtis, C.R., J.M. Scarlett, H.N. Erb and M.E. White, 1988b. Path model of individual calf risk factors for calfhood morbidity and mortality in New York Holstein herds. *Prev. Vet. Med.* **6**:43-62.

Curtis, C.R., M.E. White and H.N. Erb, 1989. Effects of calfhood morbidity on long-term survival in the New York Holstein herds. *Prev. Vet. Med.* **7**:173-186.

DeNise, S.K., J.D. Robison, G.H. Stott and D.V. Armstrong, 1989. Effects of passive immunity on subsequent production in dairy heifers. *J. Dairy Sci.* **72**:552-554.

Donovan, G.A., L. Badinga, R.J. Collier, C.J. Wilcox and R.K. Braun. 1986. Factors influencing passive transfer in dairy calves. *J. Dairy Sci.* **69**:754-759.

Duhamel, G.E. and B.I. Osburn, 1984. Neonatal immunity in cattle. *Bovine Pract.* **19**:71-78.

Foley, J.A. and D.E. Otterby, 1978. Availability, storage, treatment, composition, and feeding value of surplus colostrum: A review. *J. Dairy Sci.* **61**:1033-1060.

Gay, C.C., T.E. Besser and L.C. Pritchett, 1988. Avoidance of passive transfer failure in calves. *Proc. Am. Assoc. Bov. Pract.* **20**:118-120.

Gay, C.C., T.C. McGuire and S.M. Parish, 1983.Seasonal variation in passive transfer of immunoglobulin G_1 to newborn calves. *J. Am. Vet. Med. Assoc.* **183**:566-568.

Goodger, W.J. and E.M. Theodore, 1986. Calf management practices and health management decisions on large dairies. *J. Dairy Sci.* **69**:580-590.

Gusbi, A.M. and D.W. Hird, 1983.Calf mortality rates on five Libyan dairy stations, 1976--1980. *Prev. Vet. Med.* **1**:105-114.

Hancock, D., 1983. Epidemiologic diagnosis of neonatal diarrhea in dairy calves. *Proc. Am. Assoc. Bov. Pract.* **15**:16-22.

Hancock, D., 1985. Production symposium: Immunological development of the calf. Assessing efficiency of passive immune transfer in dairy herds. *J. Dairy Sci.* **68**:163-183.

Heinrichs, A.J. and G.L. Hargrove, 1987a. Standards of weight and height for Holstein heifers. *J. Dairy Sci.* **70**:653-660.

Heinrichs, A.J., N.E. Kiernan, R.E. Graves and L.J. Hutchinson, 1987b. Survey of calf and heifer management practices in Pennsylvania dairy herds. *J. Dairy Sci.* **70**:896-904.

Heinrichs, A.J., C.M. Jones and B.S. Heinrichs, 2003. Effect of mannan oligosaccharide or antibiotics in neonatal dairy calf diets on health and growth. *J. Dairy Sci.* **86**:4064-4069.

James, R.E., M.L. McGilliard and Hartman, D.A., 1984. Calf mortality in Virginia dairy herd improvement herds. *J. Dairy Sci.* **67**:908-911.

Kertz, A.F., 1977. Calf health, performance, and experimental results under a commercial research facility and program. *J. Dairy Sci.* **60**:1006-1015.

Losinger, W.C. and A.J. Heinrichs, 1996. Management practices associated with high death levels due to respiratory problems among preweaned dairy heifers. *JAVMA* **209**:1756-1759.

Losinger, W.C. and A.J. Heinrichs, 1997. Management practices associated with high mortality among preweaned dairy heifers. *J. Dairy Res.* **64**:1-11.

Martinez, M.L., A.E. Freeman and P.J. Berger, 1983a. Age of dam and maternal effects on calf livability. *J. Dairy Sci.* **66**:1714-1720.

Martinez, M.L., A.E. Freeman and P.J. Berger, 1983b. Genetic relationship between calf livability and calving difficulty of Holsteins. *J. Dairy Sci.* **66**:1494-1502.

Mee, J.F., 1988. Bovine perinatal mortality and parturient problems in Irish dairy herds. J. Ir. Grassland *Anim. Prod. Assoc.* **22**:106-110.

Michanek, P., M. Ventrop and B. Westrom, 1989. Intestinal transmission of macromolecules in newborn dairy calves of different ages at first feeding. *Res. Vet. Sci.* **46**:375-379.

Odde, K.G., 1988. Survival of the neonatal calf. Factors influencing colostral and calf serum immunoglobulin levels. *Vet. Clin. North Am. Food Anim. Pract.* **4**:501-508.

Place, N.T., A.J. Heinrichs and H.N. Erb, 1998. The effects of disease, management, and nutrition on average daily gain of heifers from birth to four months of age. *J. Dairy Sci.* **81**:1004-1009.

Pohjola, S., H. Oksanen, E. Neuvonen, P. Veijalainen and K. Henriksson, 1986. Certain enteropathogens in calves of Finnish dairy herds with recurrent outbreaks of diarrhea. *Prev. Vet. Med.* 3:547-558.

Pritchard, D.G., C.A. Carpenter, M.S. Richards, J.I. Brewer and S.P. Morzaria, 1981. Effect of air filtration on respiratory disease in intensively housed veal calves. *Vet. Rec.* **109**:5-9.

Reynolds, D.J., J.H. Morgan, N. Chanter, P.W. Jones, J.G. Bridger, T.G. Debney and K.J. Bunch,1986. Microbiology of calf diarrhea in southern Britain. *Vet. Rec.* **119**:34-39.

Robison, J.D., G.H. Stott and S.K. DeNise, 1988. Effects of passive immunity on growth and survival in the dairy heifer. *J. Dairy Sci.* **71**:1283-1287.

Roy, J.H.B., 1980. Symposium: Disease prevention in calves. Factors affecting susceptibility of calves to disease. *J. Dairy Sci.* **63**:650-664.

Roy, J.H.B., 1983. Problems of calf hearing in connection with their mortality and optimal growth: A review. *Livestock Prod. Sci.* **10**:339-349.

Ruckebusch, Y., C. Dardillat and P. Guilloteau, 1983.Development of digestive functions in the newborn ruminant. *Ann. Rech. Vet.* **14**:360-374.

Sieber, M., A.E. Freeman and D.H. Kelley, 1989. Effects of body measurements and weight on calf size and calving difficulty of Holsteins. *J. Dairy Sci.* **72**:2402-2410.

Simensen, E., 1982. An epidemiological study of calf health and performance in Norwegian dairy herds. II. Factors affecting mortality. *Acta Agric. Scand.* **32**:421-427.

Simensen, E., 1986. Calf mortality: Epidemiological considerations. *World Rev. Anim. Prod.* **22**:39-43.

Snodgrass, D.R., H.R. Terzolo, D. Sherwood, I. Campbell, J.D. Menzies and B.A. Synge, 1986. Etiology of diarrhea in young calves. *Vet. Rec.* **119**:31-34.

Stott, G.H., D.B. Marx, B.E. Menefee and G.T. Nightengale, 1979. Colostral immunoglobulin transfer in calves. I. Period of absorption. J. Dairy Sci. 62:1632-1638.

Stott, G.H. and A. Fellah, 1983. Colostral immunoglobulin absorption linearly related to concentration for calves. *J. Dairy Sci.* **66**:1319-1328.

Thompson, J.R. and J.E.O. Rege, 1984. Influences of dam on calving difficulty and early calf mortality. *J. Dairy Sci.* **67**:847-853.

Tozer, P.R. and A.J. Heinrichs, 2001. What affects the costs of raising dairy heifers: A multiple component analysis. *J. Dairy Sci.* **84**:1836-1844.

Tzipori, S., 1985. The relative importance of enteric pathogens affecting neonates of domestic animals. *Adv. Vet. Sci. Comp. Med.* **29**:103-206.

United States Department of Agriculture, 2002. Part I: Reference of dairy health and management in the United States.

United States Department of Agriculture, 2003. Part II: Changes in the United States dairy industry, 1991-2002.

Waltner-Toews, D., S.W. Martin and A.H. Meek, 1986a. An epidemiological study of selected calf pathogens on Holstein dairy farms in southwestern Ontario. *Can. J. Vet. Res.* **50**:307-313.

Waltner-Toews, D., S.W. Martin and A.H. Meek, 1986b. Dairy calf management, morbidity and mortality in Ontario Holstein herds. II. Age and seasonal patterns. *Prev. Vet. Med.* **4**:125-135.

Waltner-Toews, D., S.W. Martin and A.H. Meek, 1986c. Dairy calf management, morbidity and mortality in Ontario Holstein herds. III. Association of management with morbidity. *Prev. Vet. Med.* **4**:137-158.

Waltner-Toews, D., S.W. Martin and A.H. Meek, 1986d. Dairy calf management, morbidity and mortality in Ontario Holstein herds. IV. Association of management with mortality. *Prev. Vet. Med.* **4**:159-171.

Waltner-Toews, D., S.W. Martin and A.H. Meek, 1986e. The effect of early calfhood status on survivorship and age at first calving. *Can. J. Vet. Res.* **50**:314-317.

Waltner-Toews, D., S.W. Martin, A.H. Meek and I. McMillan, 1986f. Dairy calf management, morbidity and mortality in Ontario Holstein herds. I. The data. *Prev. Vet. Med.* **4**:103-124.

Webster, A.J.F., 1984. *Calf Husbandry, Health and Welfare.* Boulder, CO, Westview Press, pp. 1-202.

Weller, J.I., I. Misztal and D. Gianola, 1988. Genetic analysis of dystocia and calf mortality in Israeli Holsteins by threshold and linear models. *J. Dairy Sci.* **71**:2491-2501.

Building immunity in dairy calves

Sharon T. Franklin
Franklin Consulting, Lexington, Kentucky

1. Introduction

The immune system that protects the body from infection is composed of two main branches, the innate branch and the adaptive branch. The innate branch provides the first line of defence against an attack by infectious agents and includes such components as skin, mucous membranes, and neutrophils. The adaptive branch includes cells that can 'adapt' and specifically target many different invaders. In many respects, an attack against the body by disease causing organisms can be compared to the attack of an army against a fort. The fort and its defenders can be compared to the body and the immune system. The walls of the fort are similar to the skin and mucous membranes of the body. They are physical barriers against an invasion. The neutrophils are similar to scouting parties that are patrolling the perimeter and have orders to engage any enemy they find. They are effective killers and can also send an alarm. The commanding officers of the fort are similar to the monocyte/macrophage cells of the adaptive branch of the immune system. These cells take in much information about the invaders and send signals to other parts of the adaptive branch of the immune system about the best way to fight. The lieutenants and sergeants of the defending forces are similar to the T-cells of the adaptive branch. These cells convey the messages of the commanding officers to the troops defending the walls but they also have their own way of fighting as well. The troops lining the walls can be compared to B-cells in the adaptive branch of the immune system. B-cells are responsible for producing antibodies (or immunoglobulins, Ig) to be used in the fight against invading organisms. The antibodies can be compared to the bullets fired by the defenders in an attempt to repulse the invading army. If reinforcements had started for the fort at the first word of an impending attack, this would have been similar to the response of the immune system to a vaccination or previous exposure. With adequate reinforcements, the fort probably could hold against an invasion. Without reinforcements, the fort may be overwhelmed. When disease-causing organisms overwhelm the defences of the immune system, an animal becomes sick. Eventually, perhaps with the aid of artillery (antibiotics), reinforcements may arrive to retake the fort and the disease may be banished from the land. Adding nutrition or nutraceuticals to the picture, may make the components of the immune system stronger, faster, and more prolific. Although it may not be the deciding factor in this scenario, it may bolster the strength of the fort and its defenders so they could hold the fort until the reinforcements and artillery arrive.

2. Colostrum feeding and passive transfer of immunity

Calves are born with a naïve immune system that must be educated before it can mount an effective immune response against most pathogens on its own and generate its own reinforcements. Whereas the innate branch of the immune system is functional, the adaptive branch must expand and learn the types of invaders present before it can mount an effective defence. In essence the immune system of a newborn calf lacks bullets (Ig). For approximately the first three weeks of its life, the calf relies primarily on the bullets (Ig) acquired from colostrum to protect it against invaders. During the first three weeks of life, concentrations of Ig in the blood gradually decrease from the 24 hour value while there is an accompanying increase in numbers and percentages of B-cells (Franklin *et al.*, 1998; Nagahata *et al.*, 1991). After three weeks of age, Ig concentrations gradually increase as the immune system of the calf begins to provide its own bullets.

The health, performance, and survival of calves, especially during the first three weeks of life, rely on achieving high concentrations of Ig (antibodies) in the blood (Besser *et al.*, 1991). Without question, the successful transfer of Ig from the cow (through colostrum) to the calf is an extremely important component for minimising death losses of calves (Wells *et al.*, 1996). The method used to supply colostrum to a calf can have a major impact on absorption of Ig and the concentration of Ig achieved in the blood. The majority of studies indicate that calves allowed to nurse their dam for their first colostrum intake may not achieve adequate transfer of passive immunity. Logan *et al.* (1981) compared serum Ig concentrations of calves that suckled only with those of calves that were left with the dam but were fed approximately 1 L of hand-milked colostrum by bottle for the first feeding. Results indicated that only 23.2% of the calves allowed to suckle naturally acquired sufficient Ig whereas 42.3% of calves fed 1 L of colostrum by hand and also were left with the dam to suckle acquired sufficient serum Ig concentrations. Nocek *et al.* (1984) reported that Holstein calves fed a total of 5.45 L of colostrum in three feedings by bottle during the first 12 h after birth had higher serum protein and serum IgG concentrations than calves that suckled. Besser *et al.* (1991) reported that failure of passive transfer (serum protein < 5.0 g/dl or IgG < 1000 mg/dl) occurred in 61% of Holstein calves allowed to suckle at birth compared to 19% failure of passive transfer in calves fed 1.9 L of colostrum by bottle. Further, only 11% of calves experienced failure of passive transfer when fed 2.84 L of colostrum at birth.

We conducted a study utilising 31 Holsteins calves that were either removed at birth and hand-fed colostrum from the dam or remained with the dam and allowed to nurse (Franklin *et al.*, 2003). Hand-fed calves were fed 2.84 L of colostrum at birth and 1.9 L at 12 hours after birth. Serum protein concentrations at 24 hours after birth were greater in calves fed colostrum by hand even though calves allowed to nurse the dams were helped to nurse if they were observed to be having problems. One of the calves that was supposed to receive its colostrum by nursing its dam died during the study. The calf experienced failure of passive transfer of immunity as indicated by serum protein concentrations that did not increase above levels observed at birth.

3. Colostrum quality

One of the problems with allowing calves to nurse for their first meal of colostrum is that the quality of the colostrum is unknown. In one study, we administered oxytocin as soon after calving as possible and milked the cow with a portable milker in an attempt to collect all the colostrum produced by the cow (Franklin *et al.*, 2004). We found that of 50 cows collected for the study, four of the cows (8%) produced an inadequate volume of colostrum to feed their calf two quarts of colostrum for the first feeding. Some produced no colostrum at all. We also found that a total of nine of the 46 remaining cows had very low colostrum quality ($\leq$ 50 mg/ml) as estimated by a colostrometer. Therefore, if the calves had been allowed to nurse their dam for their first feeding of colostrum, 26% would not have received either adequate volume or quality of colostrum.

We found that colostrum quality declines during the summer at the University of Kentucky dairy farm. In the previous study, ten of eleven cows with colostrometer values at $\leq$ 60 mg/ml calved between May 1 and July 31. We also collected colostrum for a subsequent study. We began collecting colostrum in December of 2003, and continued through the end of 2004. Combining the two studies and comparing only the winter and summer samples, we found that the winter colostrum quality averaged 105 mg/ml (37 samples) compared with summer quality at 62 mg/ml (26 samples). During the summer of 2004 alone, six cows produced colostrum that was measured at 40 mg/ml or less. We do not use colostrum with values of < 60 mg/ml. Therefore, it is extremely important to hand-feed calves colostrum that has been evaluated for quality.

The National Animal Health Monitoring System surveyed heifer calves on 1,811 farms in 28 states and found that 40% had less than optimal concentrations of IgG (< 1000 mg/dl). In the United States, the most recent national data indicate that approximately 47% of calves are allowed to remain with the dam and nurse while 53% are removed prior to nursing (NAHMS, 1996, 2003). While part of the calves that remain with the dam are hand-fed their first colostrum, 30% of all calves received their first colostrum by nursing the dam only.

The most important method for enhancing immunity of calves, therefore, is to hand feed adequate amounts (preferably at least three quarts at birth and two quarts at 12 hours) of high quality colostrum (> 60 mg/ml) from healthy cows that have been vaccinated to provide antibodies against important pathogens of calves. A little extra time and trouble at the beginning of a calf's life may either save a lot of time, trouble, and expense a little later in the calf's life or save the calf's life.

4. Vitamins and trace elements supply and calves immune function

In addition to proper feeding of colostrum at birth, several nutrients may enhance immune function of calves. Some of these include the fat-soluble vitamins A, D, and E, the water-soluble B vitamins and vitamin C, and minerals such as selenium, zinc, and copper. We were interested in determining the effects of vitamin A on health and disease resistance. One thing we learned is that if a little is good, it is not always true that a lot will be better. We supplemented groups of 16 calves with either 0, 1200 (NRC, 1989 recommended levels of vitamin A), 34,000 (approximately the amount found in many milk replacers), or 68,000 (an excessive amount) IU of vitamin A/d (Hammell *et al.*, 2000). We conducted liver biopsies to monitor vitamin A concentrations as an indication of vitamin A status. We found that calves are born with practically no vitamin A stored in their liver and apparently they like it that way. Even when we supplemented calves with 68,000 IU/d, liver stores of vitamin A did not increase to levels considered adequate until calves were approximately six weeks old. We also found that we obtained better growth rates in the calves with the 1200 IU/d supplementation rate and that plasma levels of vitamin E were decreased in calves supplemented at the two highest rates of vitamin A. A subsequent study investigated levels of supplementation between the 1,200 and 34,000 IU rates we used and showed that the most effective rate of supplementation was at 10,000 IU of vitamin A/d.

5. External temperature impact on immunity

We have conducted several studies trying to determine how ambient temperature affects the immune system of calves and whether or not nutrition plays a role. Originally, we noticed that types of white blood cells present in the blood differed in winter versus summer calves (Sorenson *et al.*, 1997). In fact, we noticed alterations in the white blood cell types when we would have a wide fluctuation in ambient temperature over a short period of time. This occurred in a project conducted while I was at South Dakota State University. While at the University of Kentucky, we continued studies regarding effects of temperature on immune function (Meek *et al.*, 2004). We monitored white blood cell types present in blood of heifers during ambient temperature fluctuations from hot to warm to cold, then back to warm. We found that the same general alterations in white blood cell types occurred in calves in Kentucky as occurred in calves in South Dakota. The alterations were not as dramatic, however. A subsequent study led us to believe that the alterations in white blood cell types were in response to a need for extra energy. If calves consumed adequate energy during cold weather, the alterations in white blood cell types did not occur.

6. Supplements to support immunity

Another study we conducted in South Dakota investigated the use of supplements as support for the immune system (Donovan *et al.*, 2002). Antibiotics traditionally have been supplemented to calves in milk replacer to support the immune system. Calves subjected to stressful situations and supplemented with antibiotics have had improved growth rates compared with calves not supplemented with antibiotics. Many consumers, however, are concerned that use of antibiotics in livestock may result in drug resistant pathogens. Alternatives to antibiotics may provide equal performance. We supplemented calves with nutraceuticals (a combination of fructooligosaccharides, allicin, and gut-active microbes) and compared their performance to calves supplemented with antibiotics. Growth of calves supplemented with the combination of nutraceuticals was equivalent to growth of calves supplemented with antibiotics. Other studies have reported similar results for calves supplemented with fructooligosaccharides (Kaufhold *et al.*, 2000; Quigley *et al.*, 2002).

One of our recent studies evaluated the impact of supplementing cows with mannan oligosaccharide (MOS) during the close-up dry period on immunity of the cows and transfer of immunity to their calves (Franklin *et al.*, 2004). We used data from 39 cows to compare the effects of a control diet (n = 19;

14 Holsteins and 5 Jerseys) versus a diet supplemented with MOS (n = 20; 14 Holsteins and 6 Jerseys) on immune parameters of cows. Data from their calves, including one set each of Holstein and Jersey twins, were analysed for effects of the control diet (n = 19) versus the MOS diet (n = 22) on transfer of passive immunity to the calves. At 4 weeks prior to expected calving, blood samples were obtained from the cows and they were vaccinated against rotavirus. At 3 weeks before calving, cows were weighed, assigned to treatments, and moved to the close-up dry cow lot where they were fed the control diet or the control diet plus MOS. Cows were vaccinated a second time at 2 weeks before calving and blood samples were obtained through parturition. The cows were milked as soon after calving as possible. All cows received an intramuscular injection of oxytocin to facilitate complete removal of colostrum. The colostrum was weighed and the quality was estimated using a colostrometer. Colostrum samples were frozen for later analysis of Ig concentrations and rotavirus titers. Calves were separated from the cows prior to suckling. Blood samples were obtained at birth and 24 hours after the first feeding. Maternal colostrum was fed at the rate of 1.9 L for Holstein calves and 1.2 L for Jersey calves both at birth and 12 h after the first feeding. Blood samples were analysed for serum protein concentrations, serum Ig concentrations, rotavirus titers, packed cell volume, white blood cell counts, and white blood cell types.

The most important findings of this study are presented in Table 1. Feeding MOS during the close-up dry period resulted in significantly greater serum rotavirus titers in cows at calving, numerically greater titers in colostrum, and a tendency for greater serum rotavirus titers and serum protein concentrations in their calves. Measures of nonspecific immunity, such as Ig concentrations and white blood cell counts, were not affected by feeding MOS. The results of this study indicate that it is possible to enhance the immune response of cows during the dry period by feeding a nutritional supplement. The ability of MOS supplementation to enhance the rotavirus neutralisation titer in serum of cows immunised against rotavirus, together with the tendency for an enhancement of rotavirus titers in the serum of their calves, provides evidence that improved intestinal protection against rotavirus in calves may be achieved because of the potential for transfer of rotavirus antibodies from the bloodstream to the intestine (Besser *et al.*, 1988). Parreno *et al.* (2004) reported that elevated rotavirus titers in colostrum provided newborn calves with enhanced protection against rotavirus, resulting in fewer calves with symptoms and fewer days of diarrhoea compared with calves fed colostrum from cows not immunised against rotavirus.

Table 1. Effects of supplementation of close-up dry cows with mannan-oligosaccharide (MOS) on immune function and transfer of passive immunity to their calves.

	Control	MOS	Control	MOS	P values
	-4 weeks		Calving		
Serum rotavirus titers of cows	977	1,071	2,344	2,818	0.04
Colostrum rotavirus titers			21,777	26,009	n.s.
	Birth		24 hours		
Serum rotavirus titers of calves	110	126	4,677	7,244	0.08
Increase in serum protein concentrations of calves			1.4	1.8	0.08

The enhancement in the immune response of cows to rotavirus immunisation is an indication that responses to other pathogens of economic importance may also be enhanced through supplementation of dry cows with MOS. Further studies are needed to investigate potential benefits of supplementation with nutrients during the dry period on health and disease resistance of the cows during the transition period and transfer of passive immunity to their calves.

7. Crossbreeding and immunity

We recently investigated effects of crossbreeding on the immune system of dairy calves (Ware *et al.*, 2005). All calves were fed pooled colostrum at 5% of BW at birth and again 12 hours later. Blood samples were obtained at various times through six weeks of age and analysed for white blood cell types, serum protein concentrations, and Ig concentrations. Total Ig concentrations were greater for crossbred calves compared with the mean of the purebred calves indicating improved absorption of Ig from colostrum by crossbred calves. Additionally, differences among treatments in types of white blood cells and the phagocytosis and killing ability of neutrophils were detected. It may be that one of the most effective ways for improving immunity of dairy calves will be through crossbreeding.

8. Conclusion

Finally, no supplementation can replace good management. Providing bullets to newborn calves is the most important management practice for development of immunity in dairy calves. Other management practices can help decrease stress on calves and result in improved performance. Some important management practices include maintaining a clean environment, providing clean water and starter at an early age, and decreasing stress at weaning.

References

Besser, T.E., C.C. Gay, T.C. McGuire and J.F. Evermann, 1988. Passive immunity to bovine rotavirus infection associated with transfer of serum antibody into the intestinal lumen. *J. Virology* **62**:2238-2242.

Besser, T.E., C.C. Gay and L. Pritchett, 1991. Comparison of three methods of feeding colostrum to dairy calves. *J. Amer. Vet. Med. Assoc.* **198**:419-422.

Donovan, D.C., S.T. Franklin, C.C.L. Chase and A.R. Hippen, 2002. Growth and health of Holstein calves fed milk replacers supplemented with antibiotics or Enteroguard. *J. Dairy Sci.* **85**:947-950.

Franklin, S.T., D.M. Amaral-Phillips, J.A. Jackson and A.A. Campbell, 2003. Health and performance of calves that were suckled or were hand-fed colostrum and were fed one of three physical forms of starter. *J. Dairy Sci.* **86**:2145-2153.

Franklin, S.T., M.C. Newman, K.E. Newman and K.I. Meek, 2004. Immune parameters of dry cows fed mannan oligosaccharide and subsequent transfer of immunity to calves. *J. Dairy Sci.* **88**:766-755.

Franklin, S.T., C.E. Sorenson and D.C. Hammell, 1998. Influence of vitamin A supplementation in milk on growth, health, concentrations of vitamins in plasma, and immune parameters of calves. *J. Dairy Sci.* **81**:2623-2632.

Hammell, D.C., S.T. Franklin and B.J. Nonnecke, 2000. Use of the relative dose response (RDR) assay to determine vitamin A status of calves at birth and four weeks of age. *J. Dairy Sci.* **83**:1256-1263.

Kaufhold, J., H.M. Hammon and J.W. Blum, 2000. Fructo-oligosaccharide suplementation: effects on metabolic, endocrine, and hematological traits in veal calves. *J. Vet. Med. Ser. A* **47**:17-29.

Logan, E.F., B.D. Muskett and R.J. Herron, 1981. Colostrum feeding of dairy calves. *Vet. Rec.* **108**:283-284.

Meek, K.I., S.T. Franklin, L.J. Driedger, J.A. Jackson, M.W. Schilling and M.T. Sands, 2004. Dietary conjugated linoleic acid (CLA) affects the immune system of neonatal Holstein calves. *J. Dairy Sci.* **87** (Suppl. 2).

National Animal Health Monitoring System, 1996. *Part I. Reference of 1996 Dairy Management Practices.* Ft. Collins, CO:USDA:APHIA:VS.

National Animal Health Monitoring System, 2003. *Dairy, 2002. Part II: Changes in the United States Dairy Industry, 1991-2002.* USDA:APHIS:VS,CEAH, Fort Collins, CO

National Research Council, 1989. *Nutrient Requirements of Dairy Cattle, Update 1989.* National Academy Press, Washington, D.C.

Nagahata, H., N. Kojima, I. Higashitani, H. Ogawa and H. Noda, 1991. Postnatal changes in lymphocyte function of dairy calves. *J. Vet. Med.* **38**:49-54.

Nocek, J.E., D.G. Braund and R.G. Warner, 1984. Influence of neonatal colostrum administration, immunoglobulin, and continued feeding of colostrum on calf gain, health, and serum protein. *J. Dairy Sci.* **67**:319-333.

Parreño, V., C. Bejar, A. Vagnozzi, M. Barrandeguy, V. Costantini, M.I. Craig, L. Yuan, D. Hodgins, L. Saif and F. Fernandez, 2004. Modulation by colostrum-acquired maternal antibodies of systemic and mucosal antibody responses to rotavirus in calves experimentally challenged with bovine rotavirus. *Vet. Immunol. Immunopathol.* **100**:7-24.

Quigley, J.D., III, C.J. Kost and T.A. Wolfe, 2002. Effects of spray-dried animal plasma in milk replacers or additives containing serum and oligosaccharides on growth and health of calves. *J. Dairy Sci.* **85**:413-421.

Sorenson, C.E., S.T. Franklin, D.C. Hammell, and P.D. Evenson, 1997. Temperature and season effects on proportions of circulating mononuclear cells in Holstein calves from birth through six weeks. *J. Anim. Sci.* **75** (Suppl. 1):89.

Ware, J.V., S.T. Franklin, A.J. McAllister, J.A. Jackson, K.I. Meek, and B.G. Cassell, 2005. Evaluation of immunological differences among Holstein, Jersey, and crossbred calves. *J. Dairy Sci.* (Abstr.).

Wells, S.J., D.A. Dargatz and S.L. Ott, 1996. Factors associated with mortality to 21 days of life in dairy heifers in the United States. *Prev. Vet. Med.* **29**:9-19.

Keyword index

A

absorption 79, 96, 101
acetate 12, 13
age 92
  – at first calving 92, 93
air
  – quality 98
  – temperature 99
amino acid 39, 84
  – balance 46
  – lysine 46
  – methionine 46
antibiotic 15, 85, 107, 111
  – resistance 60
antibody 57, 95, 107, 110
antimicrobial 13
antioxidant 73

B

B-cells 107
bacteria 12, 13, 28, 45, 46, 77, 97, 99
bacterial
  – growth 45
  – yield 40
bedding 58, 101
behaviour
  – eating 85
  – feeding 85
bio-based economy 27
bio-diesel 27, 29
bio-ethanol 27, 28, 29
biomass 27
birth 89
  – weight 93
body condition score 95
bodyweight 92
by-pass starch 29
by-products 30, 31, 33

C

calf 89
  – crossbreeding 113
  – health 90
  – newborn 56, 95, 101, 108, 112
  – raising programs 89
calving
  – assistance 94
carbohydrate 27, 38, 43, 45
carrier 50, 56, 58, 62
  – active 55
  – latent 55
  – passive 55
co-products 27
colostrum 79, 89, 94, 95, 96, 101, 108, 109, 112
  – quality 109, 110
conception rates 76
copper 110

D

days open 76
deficiency 69
  – sub-clinical 71
development 19
  – tissue 17
diarrhoea 51, 55, 59, 90, 91, 92, 98, 102, 112 *See also:* scours
  – risk 93
digestibility 33, 102
  – nutrient 84
digestive tract 96, 101
distiller's grains 28, 45
dystocia 94

E

economic losses 92
energy 33

Printed in the United States
by Baker & Taylor Publisher Services